AA003284

AIChE NorCal Process Development Symposium 2014

Toward Commercialization Success

Palo Alto, California, USA
1 May 2014

ISBN: 978-1-5108-1760-9

Printed from e-media with permission by:

Curran Associates, Inc.
57 Morehouse Lane
Red Hook, NY 12571

Some format issues inherent in the e-media version may also appear in this print version.

Copyright© (2014) by AIChE
All rights reserved.

Printed by Curran Associates, Inc. (2015)

For permission requests, please contact AIChE
at the address below.

AIChE
120 Wall Street, FL 23
New York, NY 10005-4020

Phone: (800) 242-4363
Fax: (203) 775-5177

www.aiche.org

Additional copies of this publication are available from:

Curran Associates, Inc.
57 Morehouse Lane
Red Hook, NY 12571 USA
Phone: 845-758-0400
Fax: 845-758-2634
Email: curran@proceedings.com
Web: www.proceedings.com

TABLE OF CONTENTS

Development and Successful Commercialization of technology :

Examples of Recent Innovations in Polyolefins Industry

Pradeep Jain
Senior Fellow

Shrikant Dhodapkar
Fellow

The Dow Chemical Company

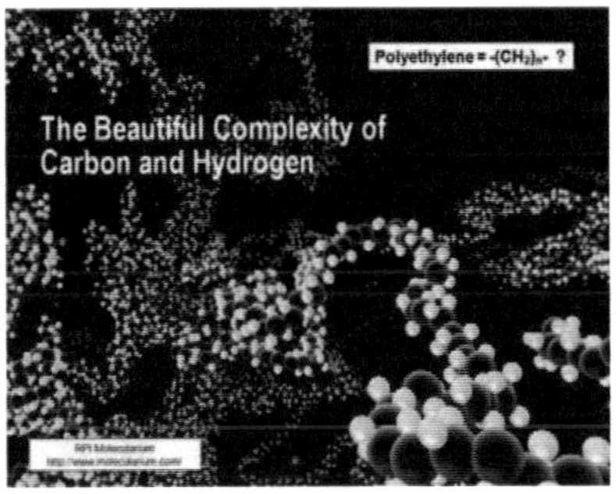

® Trademark of The Dow Chemical Company

AIChE NorCal Symposium , May 1, 2014

About Dow

- Founded in 1897 by Herbert H. Dow
- Employs more than 54,000 people globally
- Supplies more than **5,000 products** to customers in **160 countries** in sectors such as plastics, electronics, water, energy, coatings and agriculture
- From 197 manufacturing sites in 36 countries
- Generating annual sales of $60 billion
  - ✓ Largest plastic supplier to the packaging industry
  - ✓ Largest producer of polyolefin elastomers
- Dow combines the power of science and technology to passionately innovate what is essential to human progress

"If you can't do it better, why do it?"
-- Herbert H. Dow

Source: http://www.dow.com/

Outline

- Introduction to Polyolefins
  - including basic Polyethylene materials science
- Innovations brought by Molecular Catalysis
  - including metallocenes and other advanced catalysts
- Next Revolution – Olefin Block Copolymers (OBC)
- Innovations brought by OBCs
- Closing Remarks and Thoughts

Innovation Management process at Dow

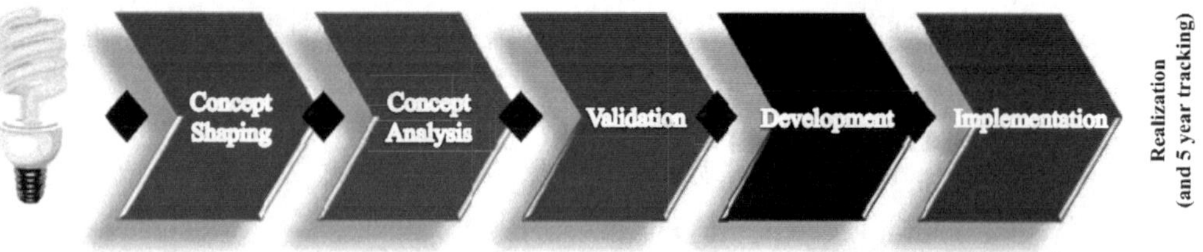

Concept Shaping	Concept Analysis	Validation	Development	Implementation
Hypothesis of Business Opportunity	Requirements for Success defined Based on VOC (Voice of Customer) analysis	Best Concept Chosen **Critical Issues Identified** Draft Business Model	**Detailed design completed and tested at Pilot-Plant level** **All issues resolved** Business plan created **Development Customer(s) is/are committed to buy**	**Implementation at Production/commercial level** Transition of the Business plan to Commercial management **Customer(s) affirms the offering, meets requirements and purchases the offering**

Polyethylene is Everywhere

- Global demand exceeds 175 Billion pounds per year!
- Growing at ~5% per year, faster growth in emerging economies

Consumer Durables and Appliances
Electrical and Electronics
Food and Specialty Packaging
Health and Hygiene
Industrial and Consumer Packaging
Pipes and Fittings
Rigid Packaging

Innovations in Polyolefins

- Polyolefins are often considered first to use in any application
  - Excellent cost / performance value
  - Ease of recycling 2 HDPE 4 LDPE 5 PP
  - Low energy to process and fabricate into films, parts, etc.

- State of the Business in 1990
  - Industry-wide overcapacity and low profitability
  - Producers discussing exit strategies such as mergers and outright sales of assets
    - Feeling that the big discoveries had already been made in this area
    - Feeling that the future belonged to those with low cost energy and feedstocks

Then everything changed!

The Evolution of Polyethylene:

An Evolution of Catalysis!

LDPE

Radical mechanism (1933)
- High Temperature & Pressure
- Many reactions possible

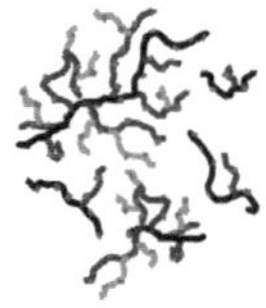

Highly Branched:
- Excellent flow properties
- Fast extrusion rates
- Poor mechanical properties

HDPE/LLDPE

Coordination catalysis (1950's)
- Low Temperature & Pressure
- Multiple catalytic sites
- Nobel Prize Ziegler & Natta 1963

Linear Backbone:
- Flexible and tough
- ***Chains are random lengths***

Metallocene PE

"Single Site" catalysts (1990's)
- Molecular catalysts
- Ability to design control into the catalyst molecule

Homogeneous Polymers:
- Chains are similar lengths
- Complete control of branching

We now have the tools to control molecular structure (and properties) with precision!

Innovation in Polyolefins

Invention + Implementation = Innovation

Control of Polymer Properties
- Molecular Weight
- Molecular Weight Distribution
- Short chain branching (determines density, crystallinity)
  - In LLDPE, SCB are incorporated by reacting with a comonomer, ex. butene, hexene or octene
- Short chain branching distribution
- Long chain branching (LCB)

Materials Science
- Structure Property Relationships
- Polymer Design
- Characterization and Methods
- Polymer Testing and Methods

Catalysis
- Reaction Chemistry
- High Throughput Screening
- Catalyst Synthesis and Testing
- Catalyst Scale-Up and Degradation
- Regulatory Testing

Process
- Reaction Kinetics and Engineering
- Process Development and Scale-Up
- Process Optimization and Implementation
- Materials Transfer and Handling
- Polymer Processing

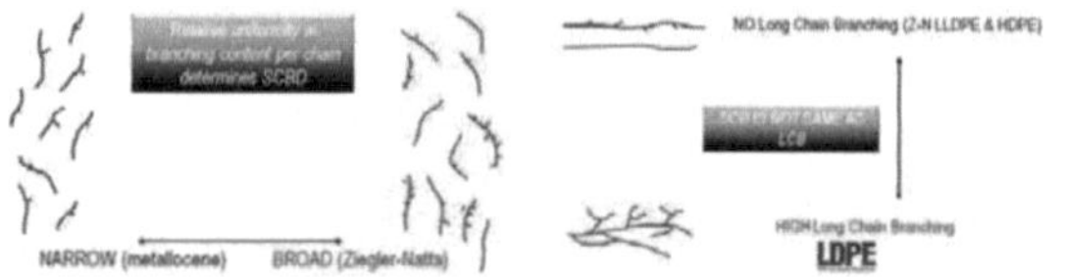

Catalysis → Reaction Engineering → Molecular Structure/Morphology → Properties

First Example of Innovation
Dow Chemical Constrained Geometry Catalyst

- >3 billion pounds/yr plastics and elastomers produced using INSITE ® catalyst
- Extremely efficient process for making elastomers
  - Less energy, less environmental impact, higher utilization of resources

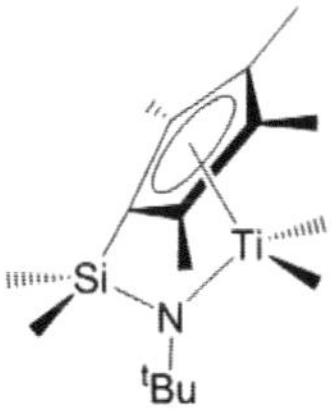

Constrained Geometry Catalyst (CGC)

® Trademark of The Dow Chemical Company

Polymer Properties Determined by Molecular Structure, Catalysis and *Kinetics*

DSC Melting of Z-N LLDPE/VLDPE and Metallocenes

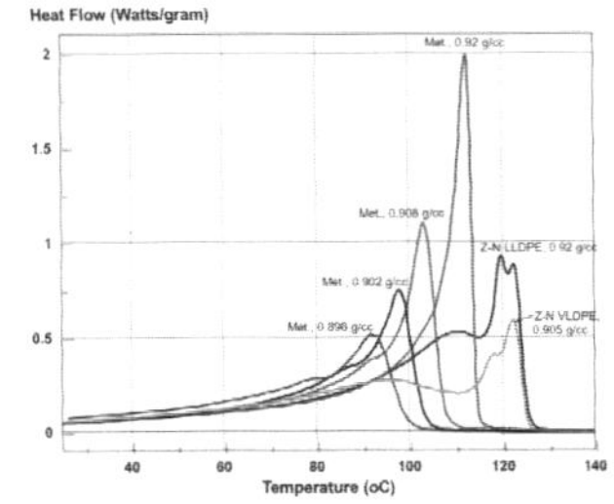

Melting Peak Temp. (°C)

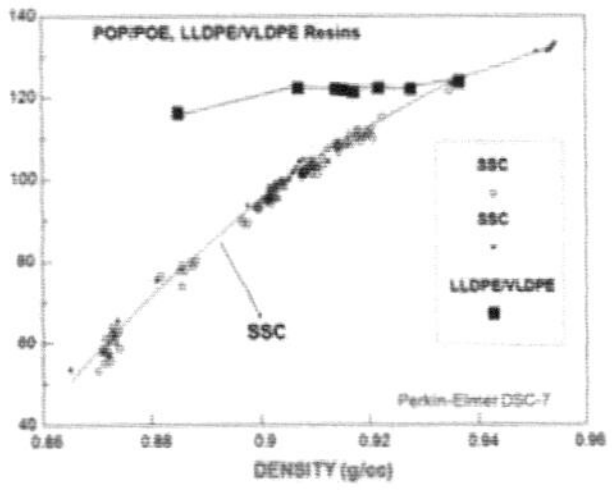

- AFFINITY ® Plastomers and ENGAGE ® Elastomers have narrow comonomer distribution
  - Better Optical Properties (Clarity, Haze & Gloss)
  - Better Elastic Properties
  - Lower T_m and Heat Seal Temperature

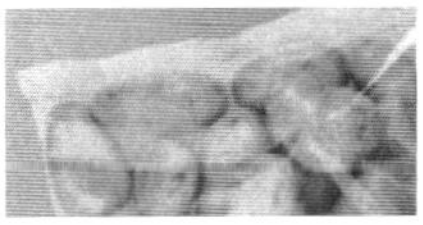

Products made with Molecular Catalysts are Narrow!

® Trademark of The Dow Chemical Company

Invention + Implementation = Innovation

Materials Science
- Structure Property Relationships
- Polymer Design
- Characterization and Methods
- Polymer Testing and Methods

Catalysis
- Reaction Chemistry
- High Throughput Screening
- Catalyst Synthesis and Testing
- Catalyst Scale-Up and Degradation
- Regulatory Testing

Process
- **Reaction Kinetics and Engineering**
- **Process Development and Scale-Up**
- **Process Optimization and Implementation**
- Materials Transfer and handling
- Polymer Processing

Innovations in Polyolefins

Kinetic Scheme and Model Development

Activation: $Cat \xrightarrow{k_a} Cat^*$

Initiation: $Cat^* + M \xrightarrow{k_i} Cat\text{-}P_1$

Propagation: $Cat\text{-}P_n + M \xrightarrow{k_p} Cat\text{-}P_{n+1}$

Deactivation: $Cat\text{-}P_n \xrightarrow{k_{da}} \text{Dead Catalyst} + D_n$

Catalyst Poisoning: $Cat^* + CP \xrightarrow{k_{cp}} \text{Dead Catalyst}$

$Cat\text{-}P_n + CP \xrightarrow{k_{cp}} \text{Dead Catalyst} + D_n$

Cat: Catalyst
Cat*: Activated Catalyst
M: Monomer
Cat-P_n: Growing Chain
D_n: Dead Chain
CP: Catalyst Poison

k_a: Catalyst Activation Rate Constant
k_i: Initiation Rate Constant
k_p: Propagation Rate Constant
k_{da}: Catalyst Deactivation Rate Constant
k_{cp}: Catalyst Poisoning Rate Constant

Propagation:

$$R_p = R_{pC_2} + R_{pC_8} = K_1[C_2][Cat] + K_2[C_8][Cat]$$

$$K_1 = \frac{k_{11}k_{22}(r_1[C_2]+[C_8])}{r_1[C_2]k_{22}+r_2[C_8]k_{11}} \quad \text{and,} \quad K_2 = \frac{k_{11}k_{22}(r_2[C_8]+[C_2])}{r_1[C_2]k_{22}+r_2[C_8]k_{11}}$$

$[C_2]$, $[C_8]$ and [Cat] are ethylene, octene and catalyst concentrations
k_{11}, k_{22} are rate constants of ethylene and octene homopolymers, respectively
r_1 and r_2 are reactivity ratios

Prediction of Chain End Vinyls:

$$\frac{\text{Vinyls}}{1000CH_2} = \left\{\frac{r_1f_1(\frac{k'_{th1}}{f} + k'_{te1}f_1 + k'_{to1}f_2)}{28(r_1f_1^2 + 5f_1f_2 + 4r_2f_2^2)}\right\} * 14{,}000$$

Prediction of Number Average Molecular Weight:

$$M_n = 28\left\{\frac{(r_1f_1^2 + 5f_1f_2 + 4r_2f_2^2)}{r_1f_1(\frac{k'_{th1}}{f} + k'_{te1}f_1 + k'_{to1}f_2) + r_2f_2(\frac{k'_{th2}}{f} + k'_{te2}f_1 + k'_{to2}f_2) + (r_1f_1 + r_2f_2)k'_{thy}\frac{[H_2]}{f}}\right\}$$

where,
M_n is number average molecular weight
f_1 and f_2 are mole fractions of ethylene and octene in reactor
f is total concentration of ethylene and octene in reactor
$[H_2]$ is hydrogen concentration in reactor
k'_{th1}, k'_{te1}, k'_{to1}, k'_{th2}, k'_{te2}, k'_{to2} and k'_{thy} are termination rate constants
r_1 and r_2 are reactivity ratios

Prediction of LCB Incorporation

$$\frac{R_{LCB}}{R_{P(C_2+C_8)}} = \frac{(1 + ar_3)}{(r_1 + a^2r_3 + 2a)}\frac{[P^=]}{[C_2]}$$

where,
r_1 and r_3 are the reactivity ratios for LCB incorporation
- Assumed the reactivity of vinyl-ended polymer chain is the same as 1-octene

$[P^=]$ and $[C_2]$ are vinyl and ethylene concentration in reactor
$a = [C_8]/[C_2]$, relative concentration of octene in reactor

Application of Model: Industrial Innovation!

- **Opportunity**: Typical Molecular catalysts are energy intensive
  - Conventional technology leads to poor productivity
- **Challenges:** High heat of reaction, poor thermal conductivity, poor mixing and fast reaction rates
- **Solution:** Design an isothermal, well-mixed reactor with good exchange capability
  - Recirculating reactor with integrated heat exchangers
  - ***US Patent 5,977,251***

United States Patent
Patent Number: 5,977,251
Date of Patent: Nov. 2, 1999
ABSTRACT

Invention + Implementation = Innovation

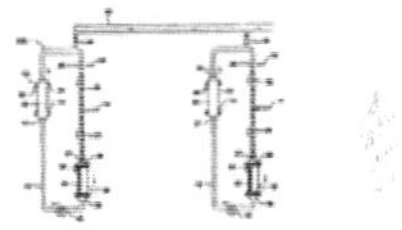

Everyday Products Using INSITE® Technology

- Car bumpers, belts, window seals, and hoses
- Athletic shoes
- "Plastic" wine corks
- Electrical wire jackets
- "Comfort-stretch" clothing
- Roofing membranes
- Food packaging

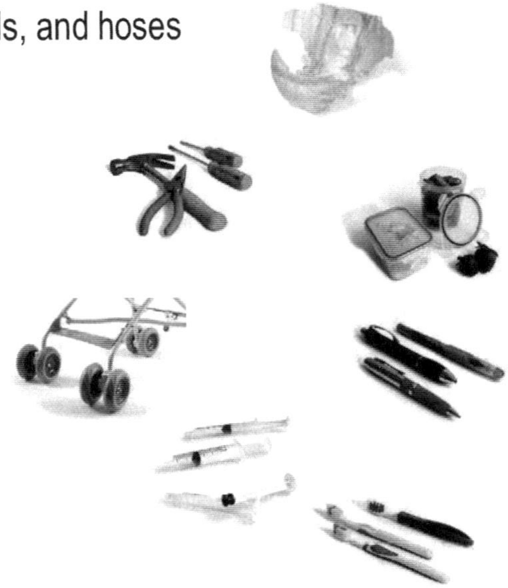

® Trademark of The Dow Chemical Company

>3 billion pounds/yr plastics and elastomers

Second Example of Innovation

Invention + Implementation = Innovation

Materials Science
- Structure Property Relationships
- Polymer Design
- Characterization and Methods
- Polymer Testing and Methods

Catalysis
- Reaction Chemistry
- High Throughput Screening
- Catalyst Synthesis and Testing
- Catalyst Scale-Up and Degradation
- Regulatory Testing

Process
- Reaction Kinetics and Engineering
- Process Development and Scale-Up
- Process Optimization and Implementation
- Materials Transfer and handling
- Polymer Processing

The Evolution of Polyethylene (Revisit)!

An Evolution of Catalysis!

LDPE

Radical mechanism (1933)
- High Temperature & Pressure
- Many reactions possible
- Kinetics complicated

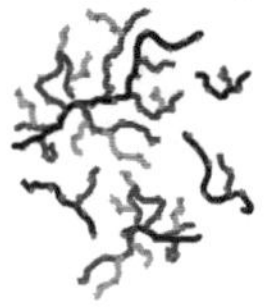

Highly Branched:
- Excellent flow properties
- Fast extrusion rates
- Poor mechanical properties

LLDPE

Coordination catalysis (1950's)
- Low Temperature & Pressure
- Multiple catalytic sites
- Nobel Prize Zielger & Natta 1963

Linear Backbone:
- PE homopolymer: crystalline
- Copolymers: flexible and tough
- Inhomogeneous

mPE

"Single Site" catalysts (1990's)
- Molecular catalysts
- Kinetics the same for each site

Homogeneous Polymers:
- Narrow molecular weight distribution
- Narrow comonomer distribution
- New monomer combinations
- Long chain branching

Finished??

Just Getting Started....New Era !!

Further Expanding the Utility of Polyolefins

soft, amorphous
ultra low density PE

hard, semicrystalline
high density PE

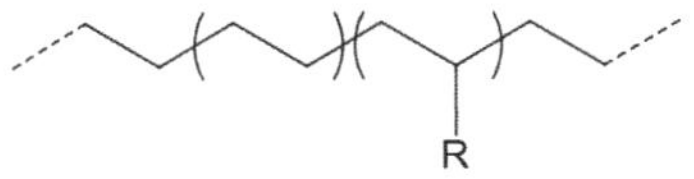

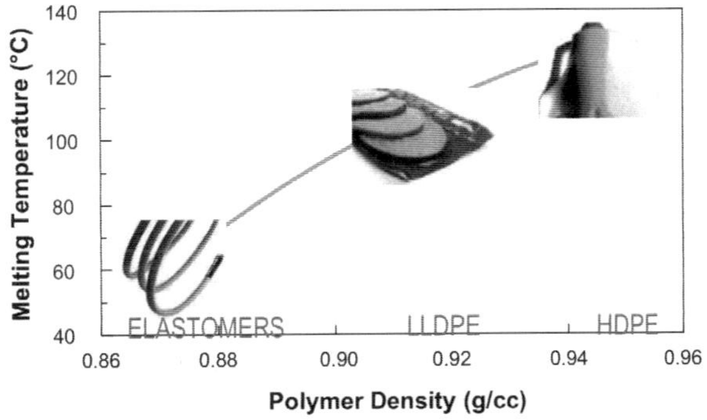

More comonomer = lower density = less crystallinity = softer material = lower melting

What if you want a crystalline elastomer with a high melting point?

Unmet Needs in Polyolefin Products: Searching for the Next Disruptive Technology

"Block" Copolymer

Pros

- Decouple properties
  - e.g. soft elastomer with high melting point

Cons

- Difficult to synthesize!!! Especially on industrial scale
  - Examples possible through living polymerization

Chain Shuttling and Catalytic OBCs

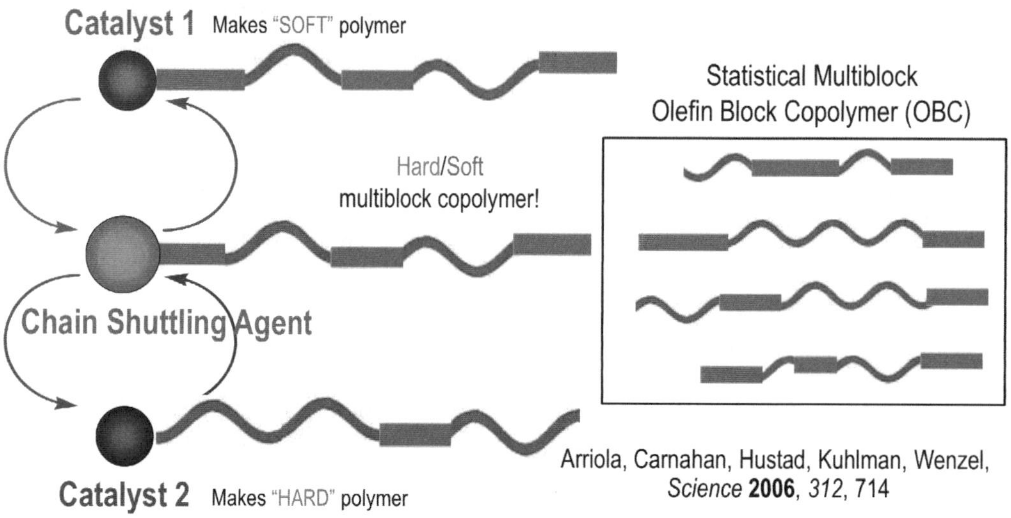

Arriola, Carnahan, Hustad, Kuhlman, Wenzel, *Science* **2006**, *312*, 714

OBCs are produced by Chain shuttling via reversible chain transfer in a continuous reactor

High Throughput Screening Identifies Suitable Catalyst/CSA Pairs for OBCs

- Screen for two catalysts of different comononer selectivities that shuttle with the same CSA

✓ Fast Shuttling rates
✓ Compatible Efficiencies
✓ Good MW response
✓ Useful density split

Good Incorporator

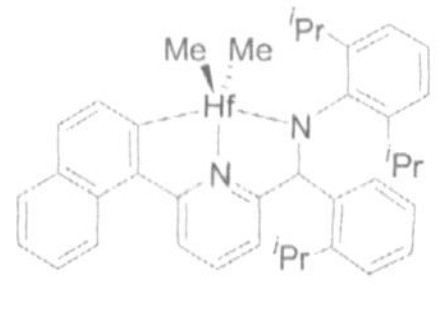

Chain Shuttling Agent

E.G. TOA, $ZnEt_2$, etc.

Poor Incorporator

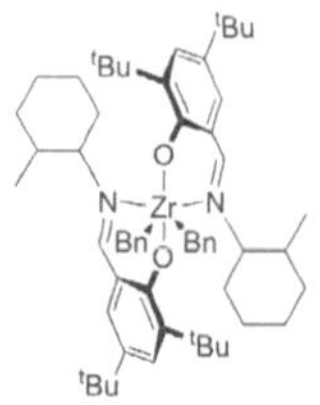

Arriola, D. J.; Carnahan, E. M.; Hustad, P. D.; Kuhlman, R. L.; Wenzel, T. T. *Science* 5 May 2006: 714.
US Patents 7,915,192 B1; 7,858,706 B1; 7,893,166 B1

Block Densities are Determined by Catalysts and Reactor Conditions

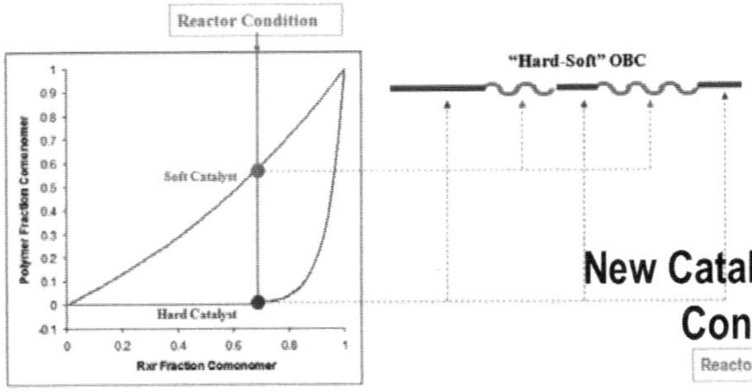

New Catalysts Can Tune Comonomer Content of the Hard Block

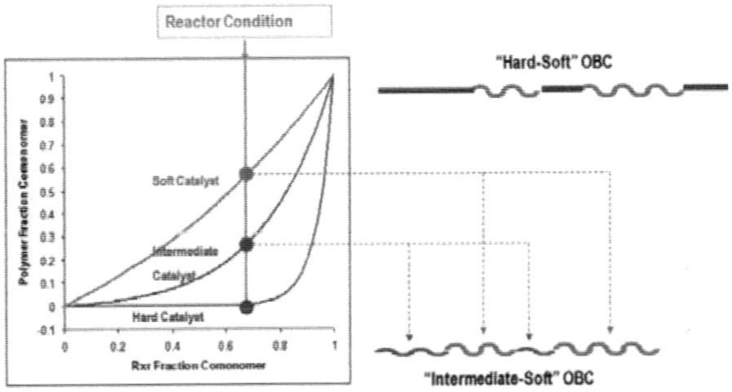

Properties Can Be Decoupled

- Ethylene-based olefin block copolymers (OBCs) from chain shuttling break many long-standing property correlations

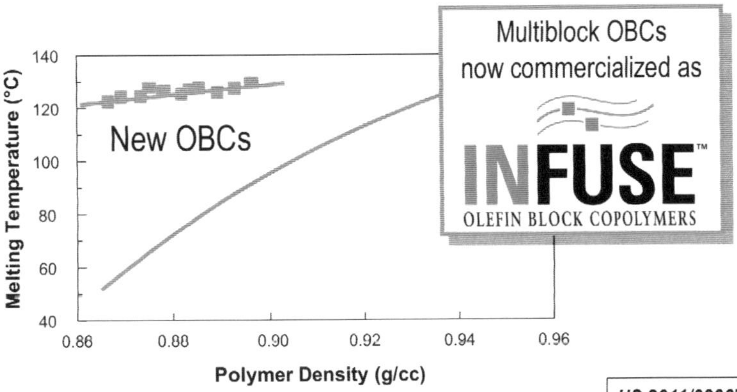

US 2011/0306715 A1

™ Trademark of The Dow Chemical Company

New Dimensions of Control

Molecular Weight Distribution

% Comonomer

Comonomer Distribution

Long-Chain Branching

← Random Copolymers

% Hard Block (split)

Difference in SCB between blocks (delta)

← INFUSE™ OLEFIN BLOCK COPOLYMERS

Block Copolymers

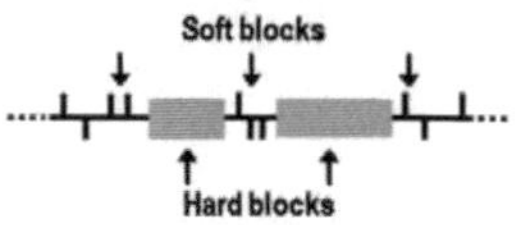

- Olefin Block Copolymers deliver flexibility AND
  - Heat Resistance
  - Compression Set
  - Elasticity
  - Processability

Random Copolymers

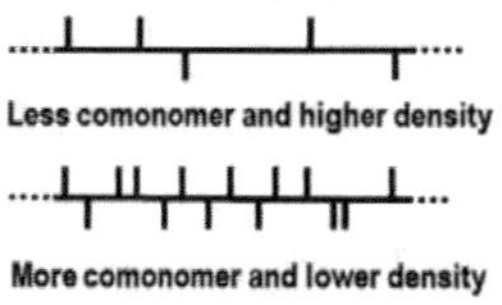

- Random Ethylene Copolymers deliver flexibility

Multiblock Ethylene-Based OBCs Commercialized by Dow as INFUSE® OBCs

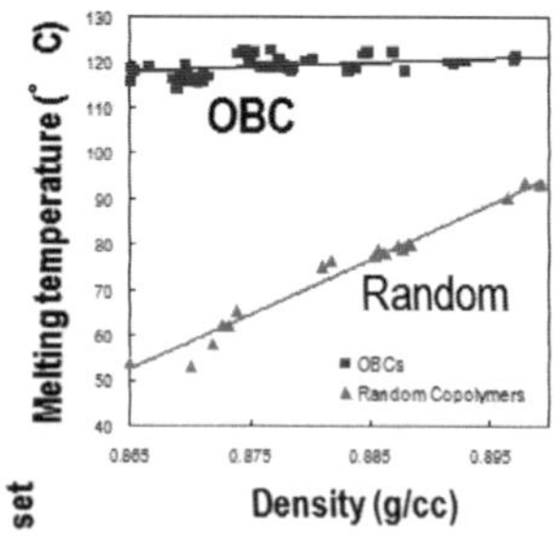

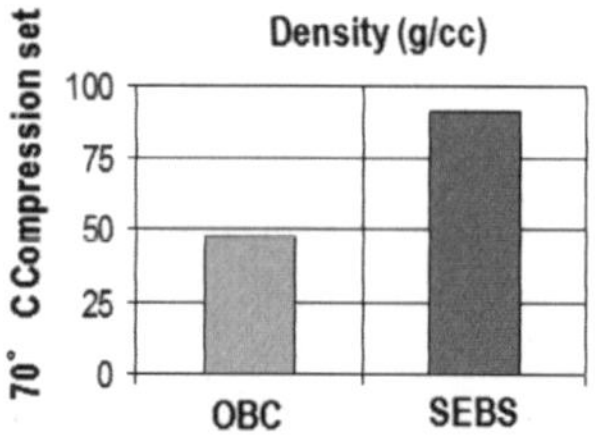

- Outstanding flexibility-temperature resistance balance
- Fast set-up in processing (shorter cycle time)
- Excellent elastic recovery properties
- Good compression set performance at room and elevated temperatures

® Trademark of The Dow Chemical Company

Multiple Market Uses

Foams

Adhesives

Building & Construction

Injection Molded Articles

Elastic films

Soft Goods and Compounds

Opportunity in Process Development
Solution Process

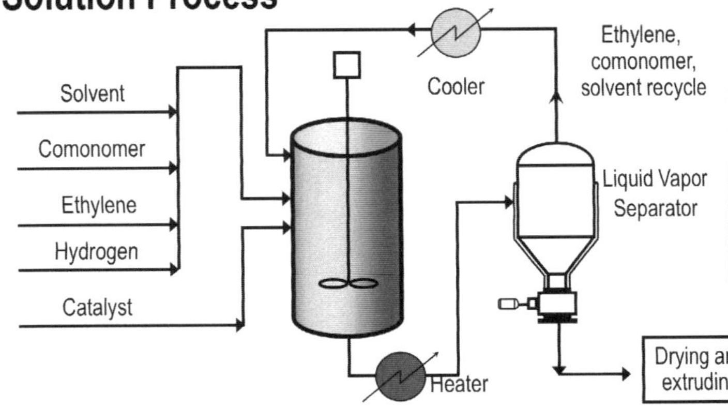

- Polymerization takes place in an inert, liquid solvent where monomer, catalyst, and polymer molecules are completely soluble
- The solvent decreases the viscosity of the reaction medium
- The solvent enables homogenous reaction conditions, in a CSTR

Opportunities in Solids Handling

Need to screen, transfer, coat, store and package pellets.
The final product is a SOLID ... and we make billions of pounds of it !!!

Slow Crystallization in Propylene/Ethylene Copolymers

The product exhibits tendency to form a compacted mass upon squeezing. After 30 minutes, the same pellets can not be squeezed into a compacted mass.

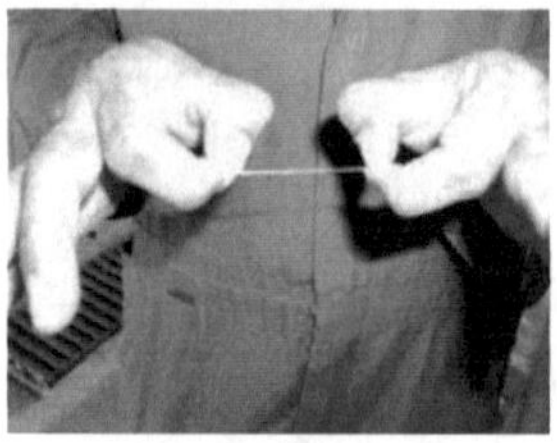

Fresh Pellets

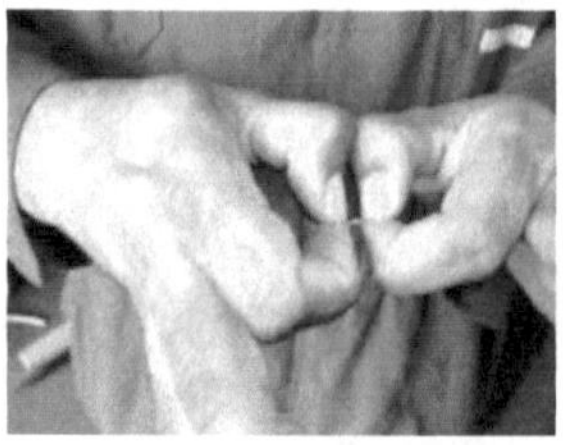

After 30 minutes

Pellets Loaded After 45 minutes

Hoppers to Create Residence Time

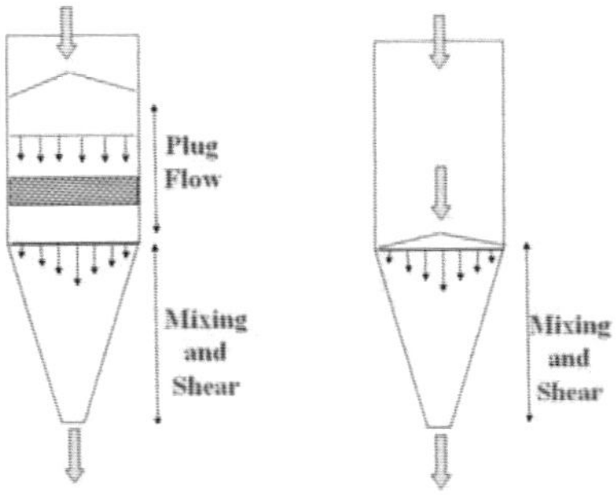

Concluding Remarks

- New catalyst discoveries are bringing new structures, *new reaction mechanisms*, new understanding in materials science and expanding into new markets
  - Random copolymers – Plastomers and Elastomers
  - Chain shuttling technology for olefin block copolymers (OBCs) with enhanced properties
- **To make money… one needs to scale-up. Successful technology development entails application of material science, catalysis and process fundamentals along with creativity !**
- Polyolefins field is thriving with opportunities for today's chemical engineer
  - Opportunities in Process Development through reduced energy dependence and process intensification
  - Opportunities in Solids Handling

Parting Thoughts

- Think ahead .... what is the rate limiting step?
- Differentiate between impossible and difficult!
- Think about unintended consequences
- Address process economics early on. Gold plated workable solutions often fail economically
- Suitable hardware may not be available - ***Embrace, enhance and adapt***
- Scalability of process equipment is critical
- **Keep your stakeholders informed and "together" at all times!**

Innovations in Polyolefins

Acknowledgements

Harold Boone
Ted Carnahan
Lisa Madenjian
Rajen Patel
Jim Stevens

The Dow Chemical Company

MANAGING PROCESS CHANGE

AICHE – May 1, 2014

Site Description

- Former plastics manufacturing and recycling facility
- Used perchloric acid in recycling activities causing soil and groundwater contamination
- Initially installed three extraction wells to control offsite migration
- Groundwater collected and disposed offsite

Groundwater Concentrations

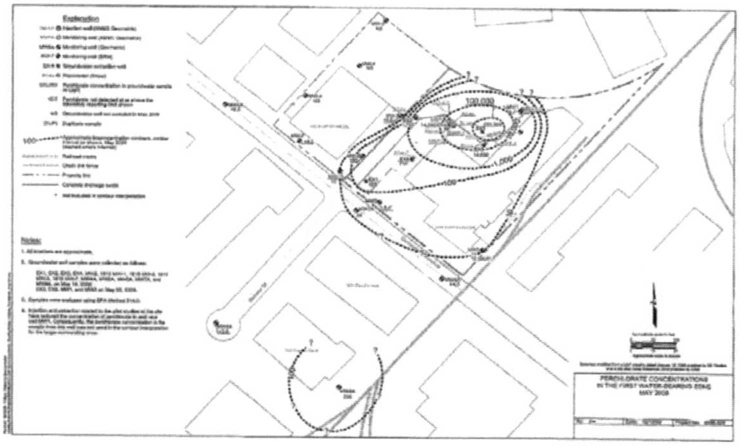

global environmental solutions

SLR

Remediation Strategy

- Control Offsite Migration and Provide Mass Removal
  - Install three additional extraction wells
    - 2 for mass removal and 1 for containment
- Remediate Source Location
  - Provide in-situ treatment by recycling treated amended groundwater using injection wells
- Discharge Excess Water – NPDES Permit

global environmental solutions

SLR

Treatment Technology Evaluation

- Disposal Cost - $800,000+ per year
- Available Technologies
  - Biological Treatment (anaerobic)
    - Packed Bed
    - Fluidized Bed
    - In-Situ
  - Physical/Chemical (adsorption)
    - Ion Exchange
    - Specialized Carbon

global environmental solutions SLR

Evaluation Criteria

- Effectiveness
- Reliability
- Complexity
- Costs
  - Capital
  - Operating
- Regulator Acceptance

global environmental solutions SLR

Design Criteria

- Expected flow rate from extraction wells 10,000 GPD
- Expected water chemistry
  - DO – 2- 4 mg/L
  - Nitrate – 10 mg/L
  - Perchlorate – 10 mg/L

global environmental solutions

SLR

Anaerobic FBR System

- Effective Proven Technology
- Very Reliable but potential upsets could occur
- Easy to Operate but More Complex than Column Systems
- Highest Capital Cost but Lowest Operating Cost
- Highest Overall Savings

global environmental solutions

SLR

Perchlorate Biodegradation

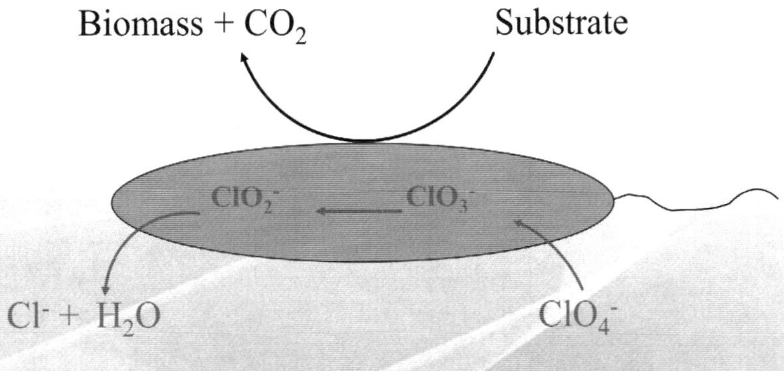

*** *Highly Favorable Reaction*** *(-801 kJ/mol acetate)*

*** *Several Different Genera Isolated*** *(Wolinella, Ideonella, Dechlorospirillum, Dechlorisoma)*

global environmental solutions

SLR

9

Utilization of Electron Acceptors

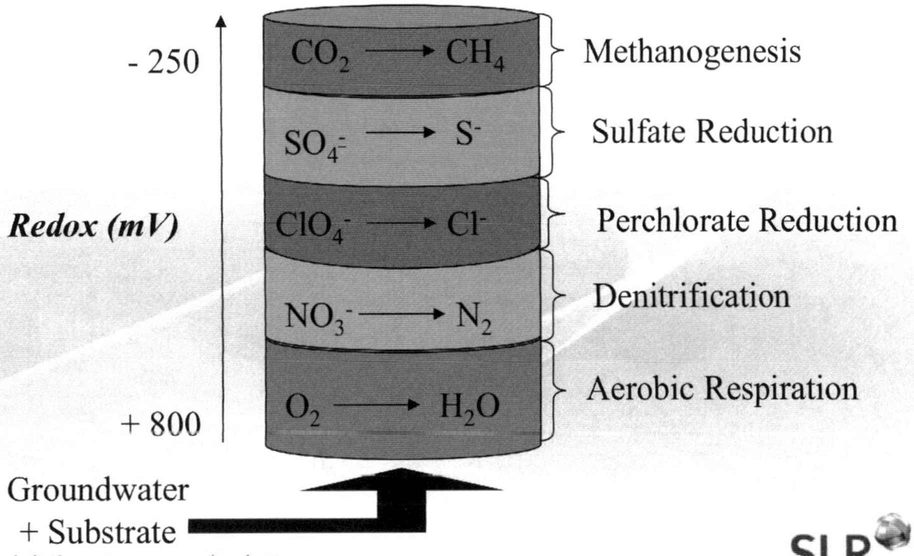

global environmental solutions

SLR

10

Fluidized Bed Reactor Flow Schematic

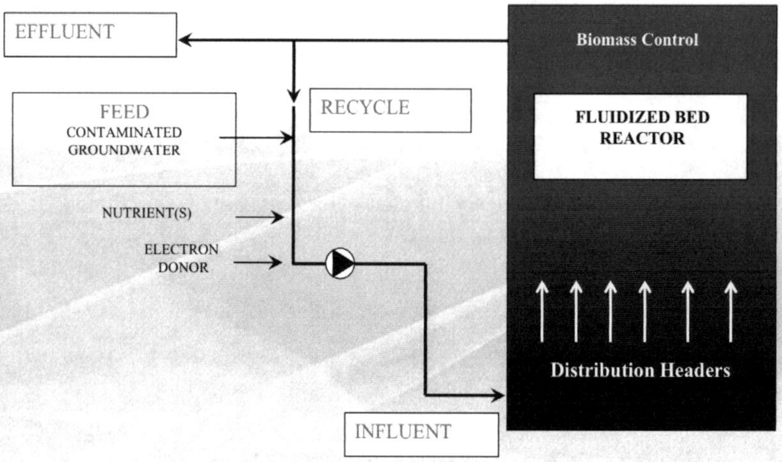

global environmental solutions

SLR

11

Site Plan

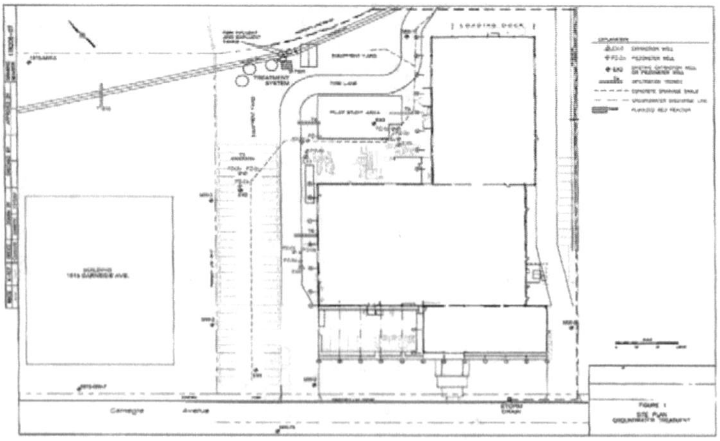

global environmental solutions

SLR

FBR Treatment System Layout

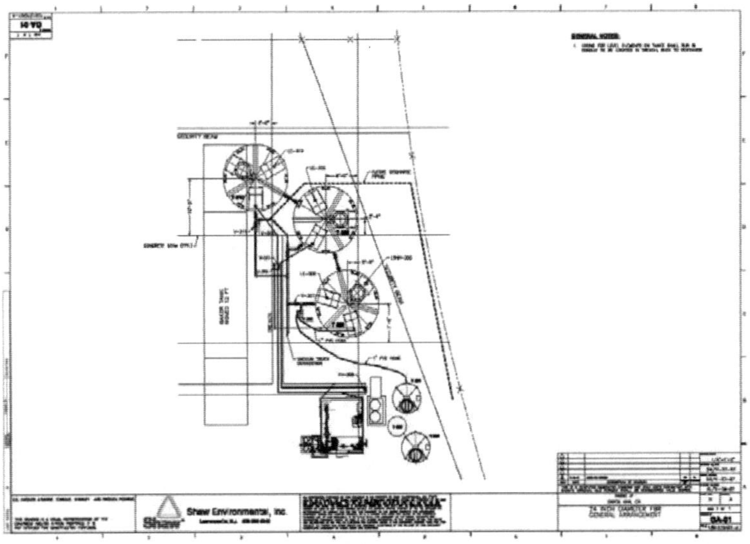

global environmental solutions

SLR

FBR Treatment System

global environmental solutions

SLR

Remediation Area

global environmental solutions

SLR

Actual Conditions

- Flow rate from extraction wells 5,000 GPD
- Water chemistry
  - DO –4 mg/L
  - Nitrate – 9-10 mg/L
  - Perchlorate – 20-30 mg/L

global environmental solutions

SLR

FBR Operating Conditions

- DO = <0.5 mg/L
- Nitrate = 0 mg/l
- Bed Expansion = 30%
- pH = 6-9 SU
- Residual Phosphorous = 0.5-1 mg/L
- ORP = -150 to -200 mv

global environmental solutions

SLR

Acclimation and Initial Results

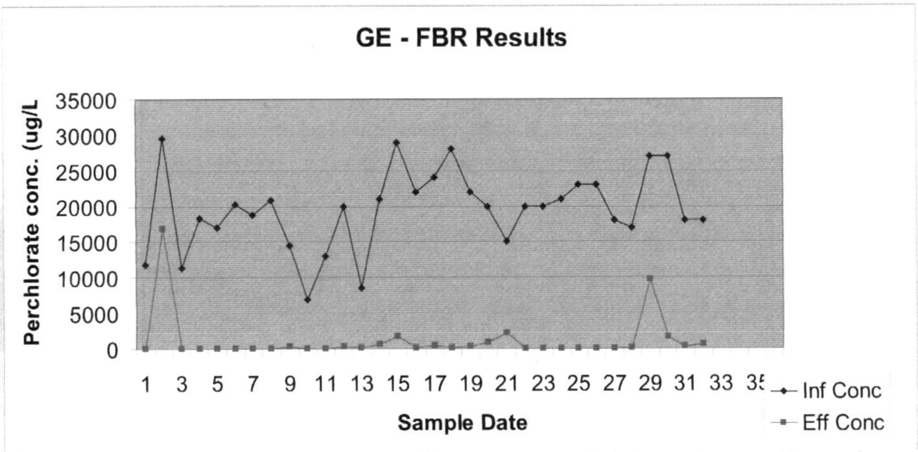

global environmental solutions

SLR

FBR System Upgrades

- Level Control on influent feed tank
- Online Instrumentation
  - DO
  - N
  - pH
  - Bed Height
- Flow meter on recycle line
- Influent Equalization

global environmental solutions

SLR

FBR Upgrades

global environmental solutions

SLR

FBR Upgrades

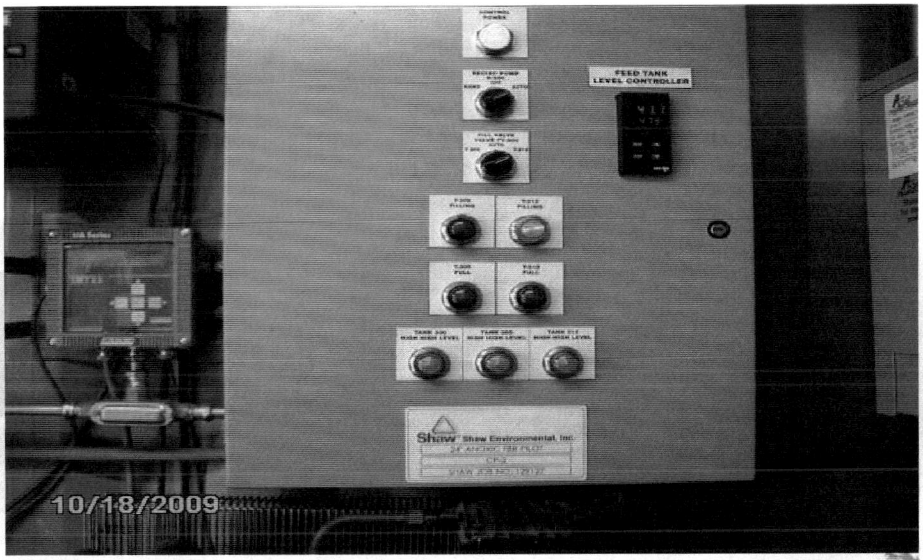

global environmental solutions SLR

Insitu Remediation Testing

- Install 4 injection wells
- Install 6 additional extraction wells
- Install 6 additional monitoring wells
- Upgrade GW treatment system to accommodate 35 GPM
- Install treated water delivery system to injection wells

global environmental solutions SLR

Regulatory Requirements

- Perchlorate Treated to < 6 ppb
- Additional Treated Water Storage of 12,000 gallons

global environmental solutions

SLR

Treatment Technology Evaluation

- Available Technologies
  - Biological Treatment (anaerobic)
    - Packed Bed
    - Fluidized Bed
  - Physical/Chemical (adsorption)
    - Ion Exchange

global environmental solutions

SLR

Lead/Lag IX System

- Effective Proven Technology
- Very Reliable
- Easy to Operate – SCADA Control
- Lowest Capital and Operating Cost
  - Low Labor requirement
  - Improved perchlorate specific resins
- Highest Overall Savings

global environmental solutions

SLR

IX Flow Diagram

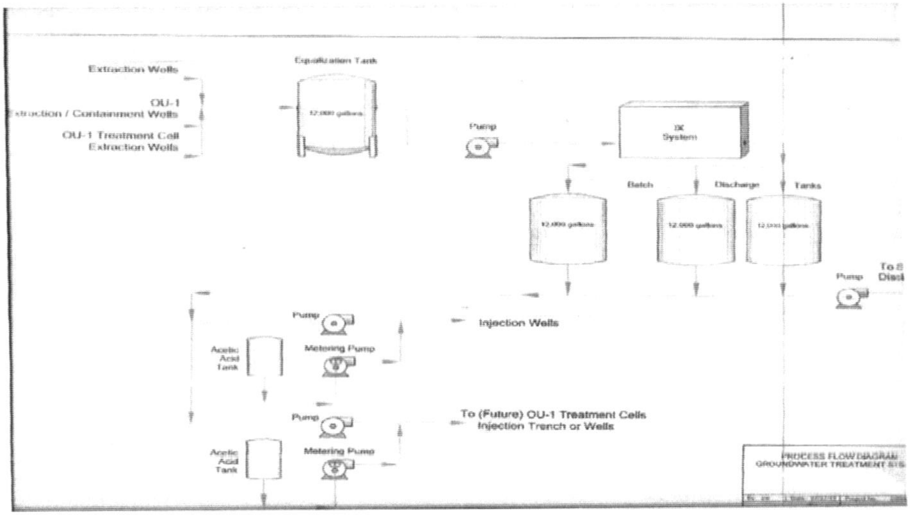

global environmental solutions

SLR

Integrated System P&ID

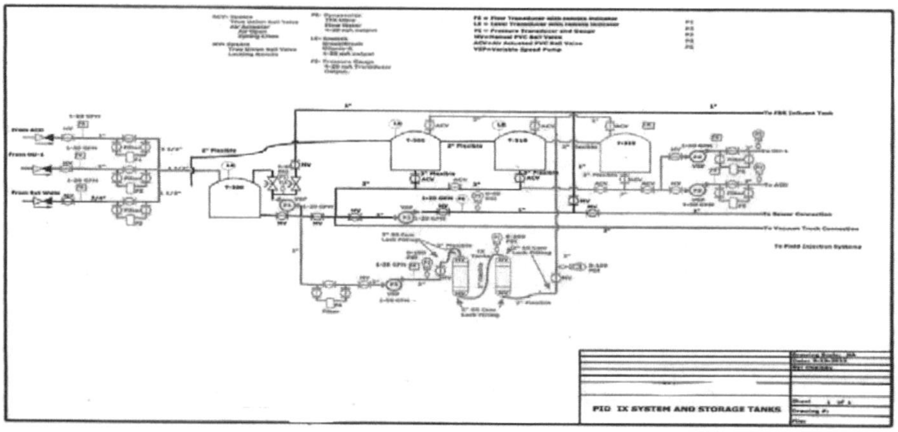

global environmental solutions

SLR

IX System Installed

global environmental solutions

SLR

Extraction Pump Controls

global environmental solutions

SLR

Instrumentation

global environmental solutions

SLR

Nitrate Analyzers

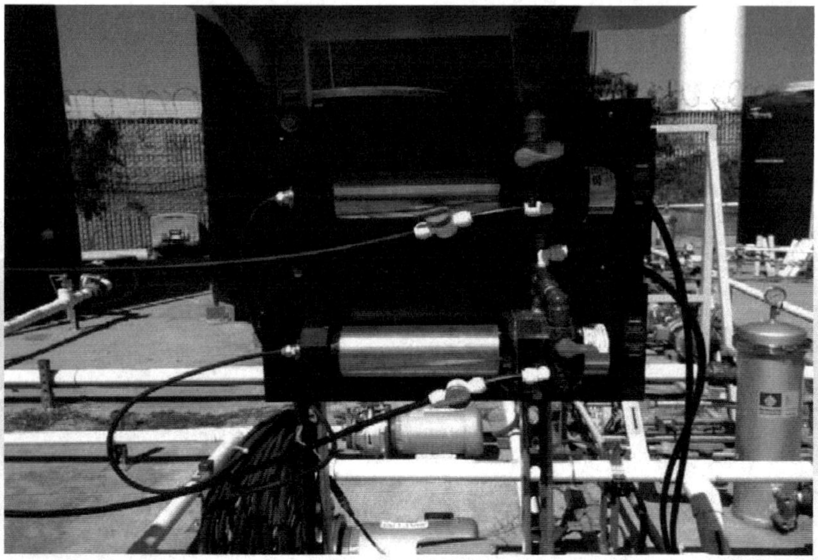

global environmental solutions

SLR

Results

- Flow Rate – 10 GPM avg.
- Perchlorate –
  - 3 ppm avg. influent
  - ND at 4 ppb effluent
- Column Capacity – 750,000 Gal. avg.
  - 60,000 to 65,000 bed volumes

global environmental solutions

SLR

SLR
global environmental solutions

QUESTIONS ?

URS | Vision to Commercialization

URS Process Technologies – Developing Technology for the Energy Industry

May 1, 2014

Agenda

1. Definitions
2. Why develop technology?
3. Technology Development in URS Process Technologies Organization
4. Case Study – CrystaSulf
5. Conclusions
6. Questions

Technology Developer vs Technology Provider

- Develop – grow or cause to grow and become more mature; advancement; start to exist.
  - A developer requires investment of personnel, money, and time
    - Requires decades of investment to be a successful developer.
    - Requires the right corporate culture and flexibility.
    - Develops higher returns but takes on more risk.
- Provide – make available for use; supply; distribute.
  - A provider requires investment of money and personnel
    - Acquires mature technology from developers.
    - Acquires key personnel for technology support.
    - Limits risk to developing the business.
- Most large engineering firms are technology providers.
- **URS is both a technology developer and a technology provider**.

Why Develop Technology?

- Differentiator in the marketplace
  - Engineering services is a commodity market
  - Provide value to both client and technology owner
  - Able to tailor technology development to industry needs or wants
- Technology evolves based on market drivers

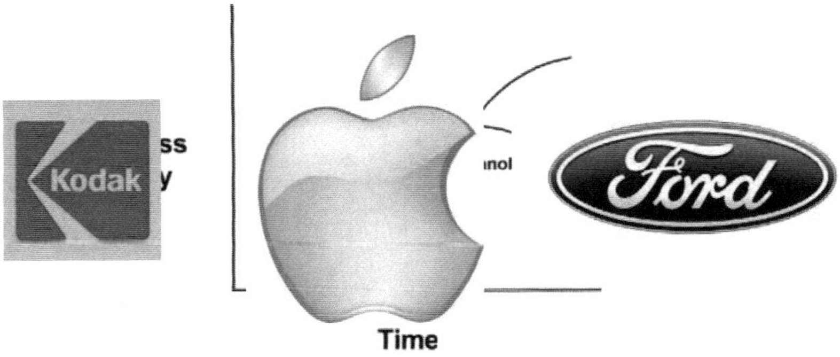

Finding and Working with Technologies

- Patience, patience, patience
  - Technology developers/inventors can make or break a deal
    - Do they understand the fair market value?
    - Can they back up their claims with statistically relevant data?
    - What value does it bring to the existing market?
  - Technology investors need to be wary shoppers
    - Understand the intended market –
      - How is it being done today?
      - What is the market size?
    - Understand the risks associated with scale-up
    - Be prepared to invest time and $
      - Customer satisfaction is independent of technology

Commercialization

Make Sure the Technology Fits

- **Understand limitations**
- **Support claims with data**
- **Seek technology protection**

- **Commit to business strategy**
- **Invest time and $**
- **Stick to known markets**
- **Develop technology support system**

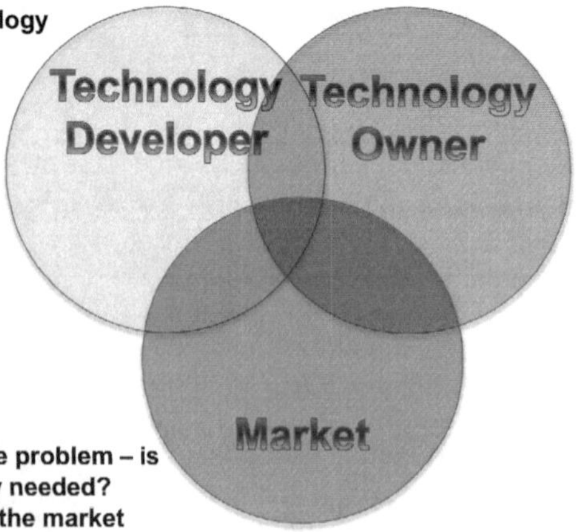

- **Understand the problem – is the technology needed?**
- **What cost will the market bear?**
- **Look beyond existing market**

URS Process Technologies

- Culture of Technology-Driven Business
  - Strong background in <u>chemistry-based</u> technologies for over 30 years
  - Emphasis in <u>developing/acquiring and commercializing technologies</u>
  - <u>Flexible</u> business model adapts to various technology positions
  - Mindset towards <u>higher risk/higher value</u> businesses and contracts
- Businesses
  - Air pollution control
  - Gas treatment

} Power/Oil & Gas/Other Applications

URS | Using Innovative Technologies   www.urs-processtechnologies.com

Flexible Format to Develop Technology

- Own technology outright: Purchase or develop technologies in-house
  - Appetite for emerging technologies that need further development either in lab or pilot scale.
  - Have <10,000 ft^2 of lab space dedicated to technology development and support.
- License technologies
  - Exclusive and non-exclusive licenses considered.
  - May have geographical limitations.
  - May be specific to a URS application.
- Develop key know-how
  - Work with technology owner during development through commercialization.
  - Intimate knowledge of technology provides value when bidding jobs using that technology.
- Maintain several technologies in different stages of development

URS | Using Innovative Technologies   www.urs-processtechnologies.com

8

CASE STUDY - CrystaSulf®

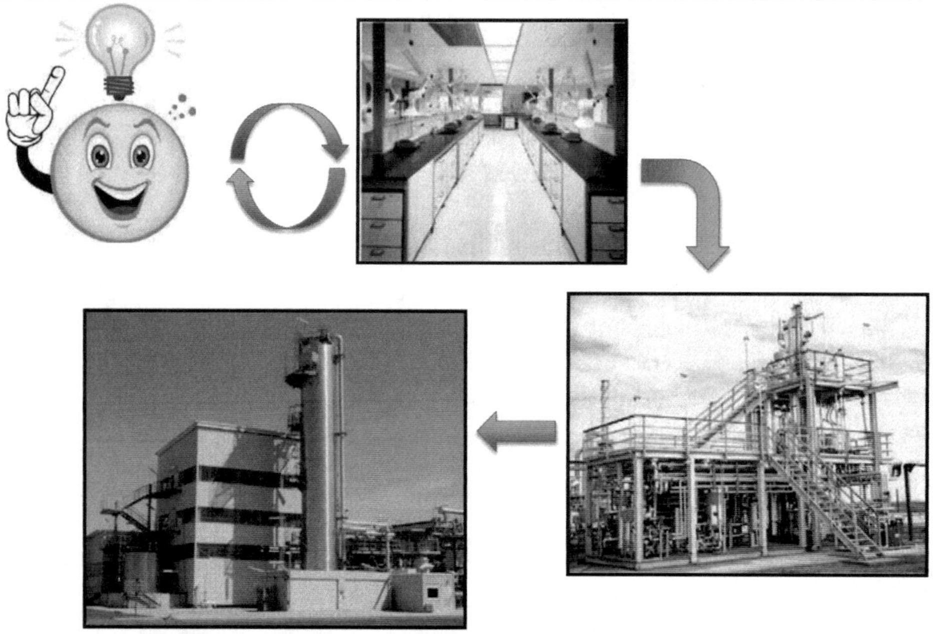

CrystaSulf Technology Development

- Why was CrystaSulf developed?
  - Gas Research Institute (GRI) industry-driven test program identified weakness in current mid-range sulfur removal technologies.
  - Industry wanted a better option.
  - Developer was selected because of intimate knowledge of the problems.

- Goals of development – the vision
  - Provide reliable high-pressure operation.
  - Minimize issues with foaming and equipment plugging.
  - Maintain or reduce operational costs.

CrystaSulf Technology Development

- How was CrystaSulf developed?
  - Started with government funded basic research
  - Bench scale was 20% co-funded by developer (labor)
  - Pilot scale was a 50:50 cost share between GRI and developer
- What was the intended market?
  - Replacement of small remotely located amine units
  - Geographic focus on TX, OK, LA region
  - Target 0.2 to 20 tons of sulfur/day

URS | Using Innovative Technologies www.urs-processtechnologies.com

11

Initial Challenge

Pilot testing successful – time to start marketing!

- Challenge: Developer ownership change
  - New ownership not interested in gas treatment technology
  - New ownership not interested in investing in new technology
- Solution: Management buyout of technology
  - Start new company (CrystaTech) w/ 7 founders
  - Have first commercial unit project in hand
  - Fortune and Fame, here we come… time to think big

Was this a solution or a bigger problem?

URS | Using Innovative Technologies www.urs-processtechnologies.com

12

The Transition

- 7 Founders = 7 Employees => 7! Opinions
- Focused on future not present
  - 2 remote sales offices
  - Spending significant $ on marketing
  - Growing the company too rapidly – adding employees

And then it happened….

- First project was canceled due to ownership change of client
- Dissention among founders – several left the company
- Realization that it was a difficult model to sustain
  - Limited marketing effort during project execution
  - Heavy reliance on outside help to fully execute

Path to Commercialization

- Persistence, Patience and Luck (not all good)
  - Broadened market to difficult to treat streams
  - Realized limitations and developed alliances and partnerships
    - Brought credibility to small company
    - Able to provide better execution services
- Sustained and grew business at CrystaTech for 12 years
  - Sold 6 technology licenses
    - 3 were fabricated
    - 3 licenses were to international clients
  - Provided other services (consulting) to augment the cash flow

The Next Step

•Sold the CrystaSulf technology to URS in mid-2011

+
- Provided large company support
- Ability to stand behind the product
- Ability to execute projects
- Established client base and networking
- **Dedicated group for technology development and deployment**

—
- Inherited large company red tape
- Increased cost of learning
- Investment in technology transfer is a slow process
- Increased cost of doing business is higher and could shift target market

The key to success is learning to adapt technology to market influences

URS | Using Innovative Technologies www.urs-processtechnologies.com

15

CrystaSulf Goals Then and Now...

•THEN

1. Replace remote amine units
2. Replace problematic technologies
3. Achieve robust operation at high pressure
4. Target TX, OK, LA sales region

•NOW

1. None - Existing infrastructure able to accommodate increasingly sour natural gas.
2. Displace portion of existing technologies
3. Exploit proven robust operation at high pressure
4. Target global market
5. Target high CO_2 applications at pressure
6. Target difficult to treat streams containing H_2S
7. Adapt technology for low pressure applications

URS | Using Innovative Technologies www.urs-processtechnologies.com

16

Conclusions

- New technologies can be acquired, developed, or licensed
  - Understand that the level of investment is directly proportional to the maturity of the technology.
  - Work with technologies that are in a familiar market.
  - Be prepared to support the technology during the early days of commercialization. Beware of the "bleeding" edge of technology.

- The technology and/or the market can be adapted
  - Be wary of myopic developers who can stall a good idea.
  - Stagger technology rollout – need a mix of emerging and established technologies.
  - Be persistent – if the technology has value, it may take the right opportunity before it can be proven.

AIChE NorCal Process Development Symposium, Palo Alto, May 1, 2014

Alternative Chemical Methods of CO_2 Mitigation

Greg H. Rau
Institute of Marine Sciences
University of California, Santa Cruz
And
Carbon Management Program
Energy and Environment Directorate
Lawrence Livermore National Laboratory

rau4@llnl.gov

Thanks to: Ken Caldeira, Bill Langer, Kevin Knauss, Susan Carroll, Roger Aines, Jeff Roberts, Nigel Quinn, and the California Energy Comm.

Why we need to consider alternative CO_2 mitigation strategies

Current technologies and policies are failing to stabilize CO_2; Poses significant, global climate and environmental risks (latest evidence: IPCC, 2014)

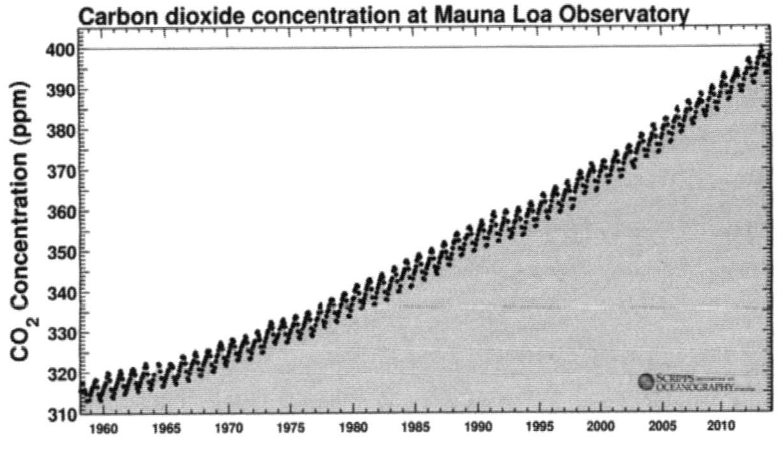

Strategies for Stabilizing/Reducing Atmospheric CO_2

- Transition to non-fossil energy use
- Increase efficiency of fossil energy use
- Fossil fuel switching, e.g. coal → nat. gas, biomass
- Pre-emissions removal of CO_2 from fossil fuel combustion,
  - CCS – Carbon Capture and Storage - the capture, concentration and underground storage of CO_2 in molecular form
  - Alternative point source mitigation approaches
- <u>Air CO_2 management</u>
  air CO_2 removal +-natural CO_2 emissions reduction

Strategies for Stabilizing/Reducing Atmospheric CO_2

- Transition to non-fossil energy use
- Increase efficiency of fossil energy use
- Fossil fuel switching, e.g. coal → nat. gas, biomass
- Pre-emissions removal of CO_2 from fossil fuel combustion,
  - CCS – Carbon Capture and Storage - the capture, concentration and underground storage of CO_2 in molecular form
  - Alternative point source mitigation approaches
- <u>Air CO_2 management</u>
  air CO_2 removal +-natural CO_2 emissions reduction

Point-Source CCS Realities

CCS can effectively mitigate CO_2 at some scale, but it's:

- ❑ Expensive
  - ➢ \$50-\$150/tonne CO_2 avoided
- ❑ Energy intensive
  - ➢ 10% to >30% energy penalty
  - ➢ decapture from absorbant and further concentration and pressurization strongly endothermic
- ❑ Difficult to retrofit
- ❑ Risky
  - ➢ conc CO_2 must be contained without leakage
    - underground storage can be problematic
    - must insure against leakage, water contamination, seismic effects, etc.
  - ➢ conc CO_2 use is limited

Alternative point-source mitigation strategies:

Examples provided by mitigation of other gaseous pollutants eg. SO_x, NO_x, NH_3, Hg, etc

- ➢ All non-CO_2 mitigated via reaction to form other compounds, or adsorption.
- ➢ No examples of capture, concentration and storage?!
- ➢ So what's is so special about CO_2 mitigation?

CO_2 is a reactive gas, e.g.:

- ➢ CO_2 + Amines ⇄ R-CO_2
- ➢ $CO_2 + NH_3 + H_2O$ ⇄ NH_4HCO_3
- ➢ CO_2 + NaOH ⇄ $NaHCO_3$
- ➢ CO_2 + MgO ⇄ $MgCO_3$

(first four reactions: **CCS Capture Apps.**)

- ➢ $CO_2 + MgSiO_3$ ⇄ $MgCO_3 + SiO_2$
- ➢ $CO_2 + CaCO_3 + H_2O$ ⇄ $Ca(HCO_3)_2$

$\Delta G^o < 0$ for all of these capture reactions

$\Delta G^o >> 0$ if reversed to recover conc. CO_2 and reagent

Chemical CO_2 Mitigation Options:

- **CCS Strategy – recycling of reagent(s):**

  waste CO_2 + reagent → carbonated intermediate

  carbonated intermediate + energy → conc. CO_2 + reagent

  issues: energy intensive, costly, risky conc. CO_2 transport and storage

- **Alternative Strategy - once-through, no recycling**

  waste CO_2 + reagent → carbonated end product (+energy)

  issues: reagent must be abundant and cheap an/or the value of the product(s) helps offsets the cost of the reagent

Alternative Strategy Example: Accelerated Weathering of Limestone, AWL

The AWL process:

$CO_{2g} + CaCO_{3s} + H_2O$ ---> $Ca(HCO_3)_{2aq}$, $\Delta G^o = -38$ kJ/mol

- CO_2 capture is exothermic and spontaneous
- $Ca(HCO_3)_{2aq}$ provides stable, benign C storage (in seawater/ brines)

Stoichiometry: (tonnes/tonne CO_2)

$(1)CO_{2g} + (2.3)CaCO_{3s} + (5000)H_2O$ ---> $(3.7)Ca(HCO_3)_{2aq} + (5000)H_2O$

Analogous to power plant FGD:

$SO_{2g} + CaCO_{3s} + 0.5O_2$ ---> $CO_{2g} + CaSO_{4s}$

Proof of Concept: California Energy Commission funded project

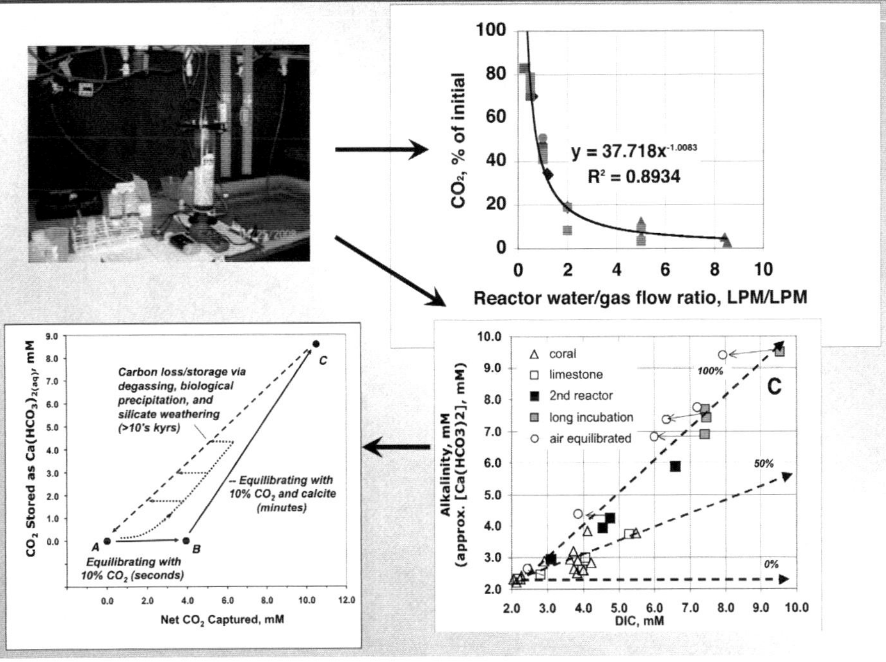

AWL Reactor Scheme

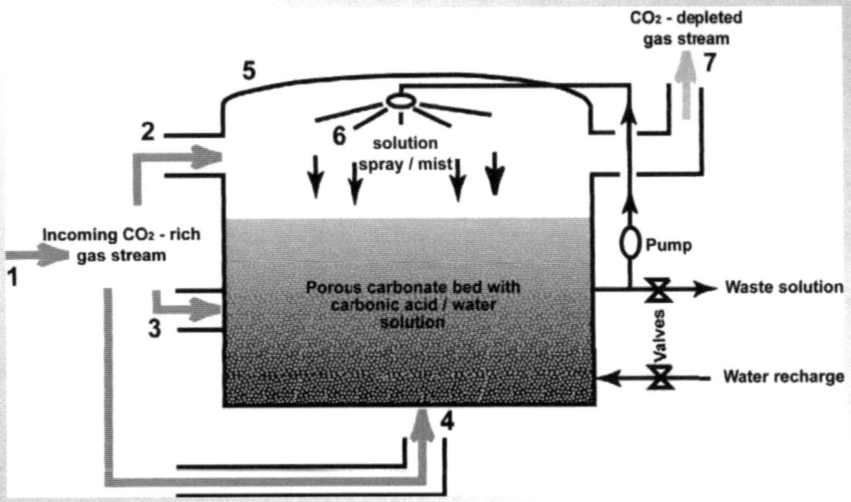

(Rau and Caldeira, 1999)

McDermott's limestone CO_2 scrubber concept

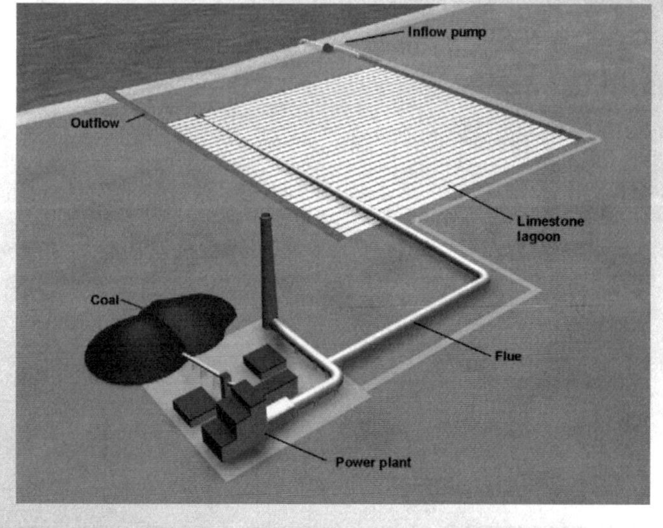

William Downs and Hamid Sarv. 2002. CO_2 CAPTURE AND SEQUESTRATION BY A LIMESTONE LAGOON SCRUBBER. McDermott Technology, Inc., Alliance, OH. 2nd Ohio CO_2 Reduction, Capture & Sequestration Forum, Ohio University, April 26 2002

Process Engineering Schematic and Issues

CO_2 (1t) → Reactor

Limestone (2.5t) → Reactor

Seawater/Brine (≈5,000t) → Reactor

Reactor → Water (≈5,000t) + Bicarbonate (3.7t)

Objective: minimize COM cost per tonne CO_2

Primary Issues:

1)Local handling of flue gas/CO_2
2)Extraction, processing, delivery, storage, handling of limestone.
3)Pumping/delivery of water.
4) Application siting that optimizes preceding.

1)Design contacting of CO_2, limestone, and water that minimizes energy and material handling costs.
2)Site application where material transport and handling costs are minimized.

Determine and minimize
1)downstream thermal, chemical and biological impacts.
2)solid waste issues
3)discharge permitting issues.

AWL Economics

Estimated cost per tonne CO_2 sequestered, assuming coastal location:

- Limestone -
  - 2.5 tonnes @ \$4/tonne = \$10.00
  - crushing from 10 cm to 1cm = \$ 1.45
  - transport 100 km by rail = \$ 8.00
- Water -
  - 5000 m^3, pumped 2 vertical meters = \$ 3.78
- Capital and maintenance = <u>\$ 2.50</u>

TOTAL: **\$ 26/tonne CO_2**

Versus to \$50-\$150/tonne for CCS

Optimum AWL Economics

Estimated cost per tonne CO_2 sequestered, assuming coastal location:

- Limestone - <u>\$/Tonnes CO_2 mitigated</u>
  - 2.5 tonnes @ \$4/tonne = ~~\$10.00~~ | *use free, nearby*
  - crushing from 10 cm to 1cm = ~~\$ 1.45~~ | *waste limestone*
  - transport 100 km by rail = ~~\$ 8.00~~ |
  - Water -
  - 5000 m^3, pumped 2 vertical meters = ~~\$ 3.78~~ *use cooling water*
- Capital and maintenance = <u>\$ 2.50</u>

TOTAL: **<\$3/tonne CO_2**

AWL advantages:

- Avoids costly/risky conc. CO_2 production and storage

  $Ca(HCO_3)_{2aq}$ addition to the ocean
  - Safely exploits vast ocean storage potential

    $Ca(HCO_3)_{2aq}$, 4th most abundant seawater compound
  - "unlimited market" for counteracting ocean acidification
- Builds on existing wet limestone scrubbing for SO_x
- Retrofitable
- Useable in developing world; this is not rocket science!
- Mimics natural global-scale CO_2 absorption process - carbonate mineral weathering.

Further Details: Rau, *ES&T,* 2011; http://pubs.acs.org/doi/pdf/10.1021/es102671x

CO_2 Mitigation Strategy #2: Non-Point Source Air CO_2 Management

- *"...the cost of this [Direct Air Capture] system is estimated to be of the order of $600 or more per metric ton of CO_2....Thus, DAC is not currently an economically viable approach to mitigating climate change."* Socolow et al. (*APS*, 2011)

- *"We estimate that total system costs of an air capture system will be on the order of $1,000 per tonne of CO_2, based on experience with as-built large-scale trace gas removal systems."*

  House et al. (*PNAS*, 2011)

However, these conclusions thankfully do not apply to air capture systems that don't make concentrated CO_2. ------->

Natural "Direct Air Capture": Globally Effective and "Free"

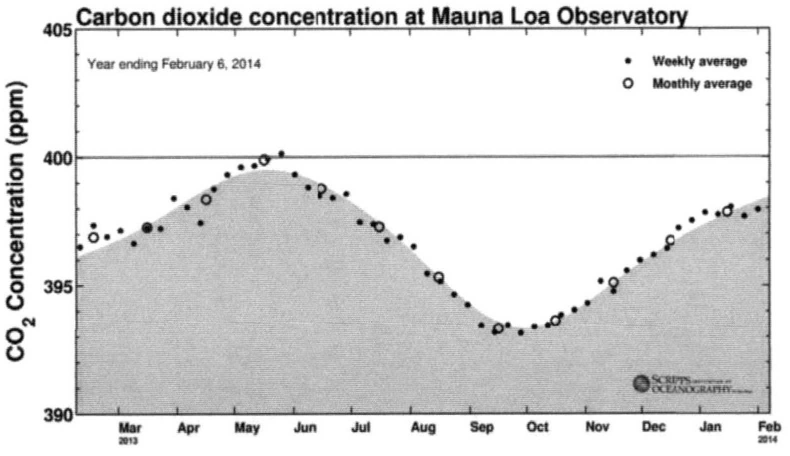

Natural, Free "Direct Air Capture" Processes:

Atmospheric CO_2

Photosynthesis

$CO_2 + H_2O + \text{photons} \longrightarrow R\text{-}(CH_2O) + O_2$

Ocean uptake

$CO_2 + H_2O + CO_3^{2-} \longrightarrow 2H^+ + 2HCO_3^-$

Weathering Reactions

Silicate weathering: $CO_2 + Mg/CaSiO_3 \longrightarrow Mg/CaCO_3 + SiO_2$

Carbonate weathering: $CO_2 + H_2O + CaCO_3 \longrightarrow Ca^{2+} + 2HCO_3^-$

Electro-Geochemical Air CO_2 Removal + C-Negative H_2 Production

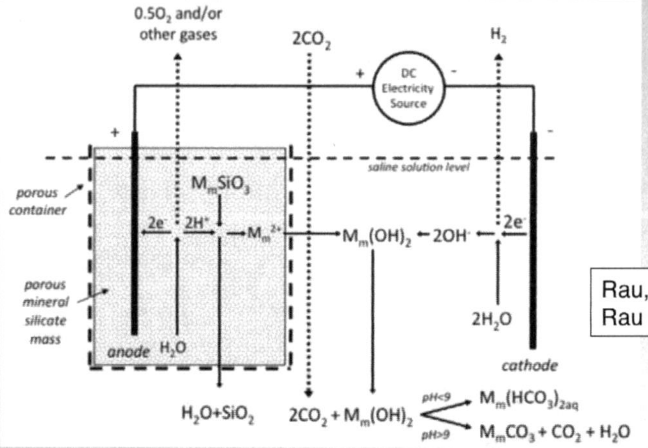

Rau, *ES&T*, 2008
Rau et al., *PNAS*, 2013

net reaction: $CO_2 + MgSiO_3 + H_2O + V_{dc} \rightarrow Mg(HCO_3)_{2aq} + H_2 + 0.5O_2$

Net cost: <\$100/tonne CO_2 mitigated

Net energy cost: <200 kJ/mol vs >400 kJ/mol for DAC

19

Experimental Setup and Results at LLNL:

→

↓

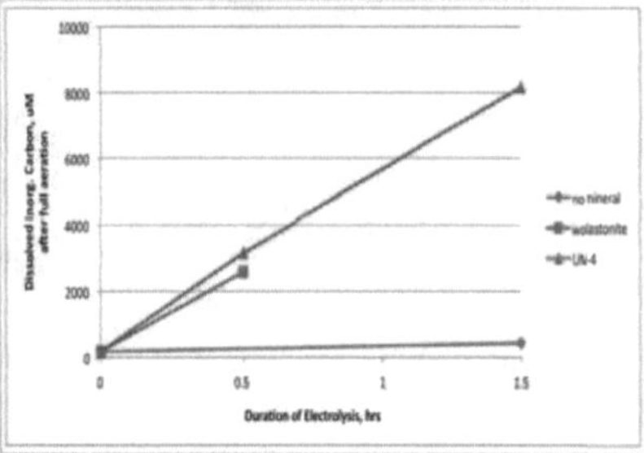

←

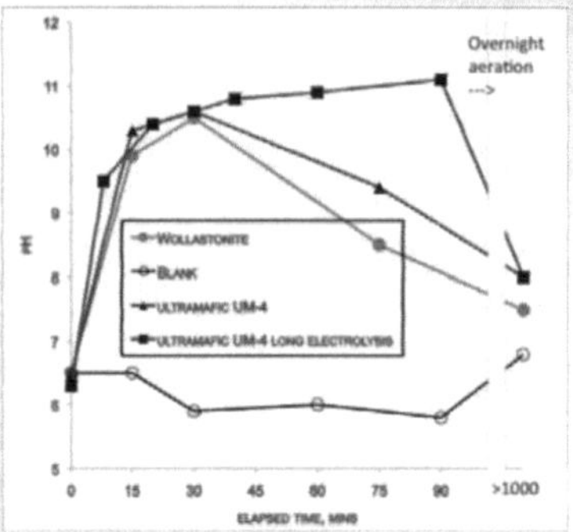

20

Implications:

Use globally abundant saline/brine water, carbonate/silicate minerals and renewable energy to: i) capture and store CO_2 from air or waste gas, and ii) produce carbon-negative H_2, all at large scales:

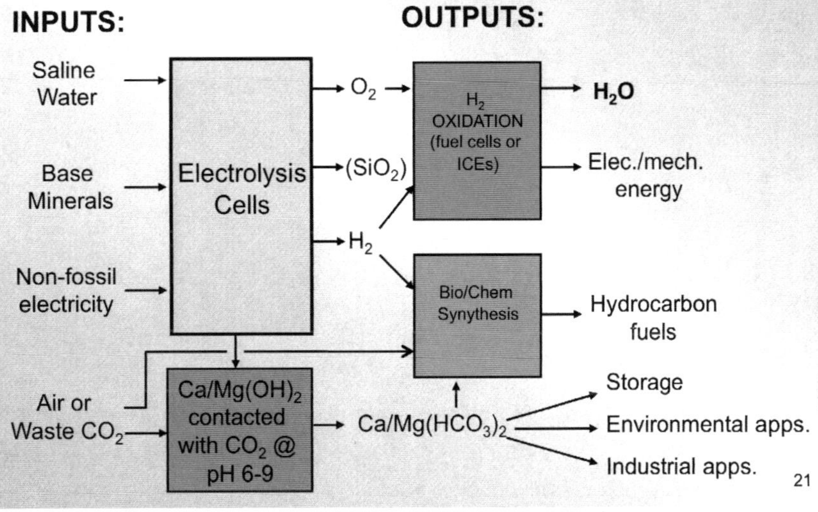

21

Energy, Reactant and Product Quantities

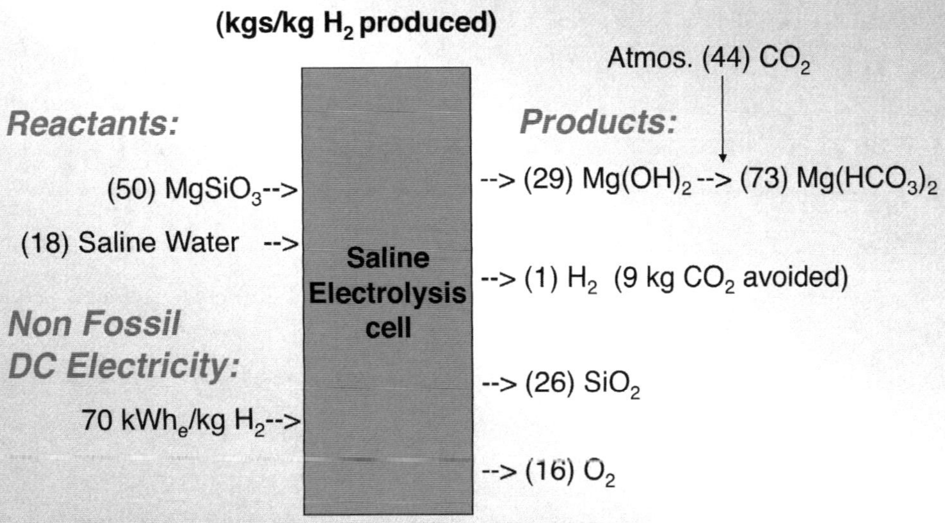

Net CO_2 mitigated: <u>53 kg CO_2 mitigated per kg H_2 generated</u>

22

Use/Fate of the Dissolved Bicarbonate Produced

Demonstrated: $CO_2 + H_2O$ + minerals -----> HCO_3^-

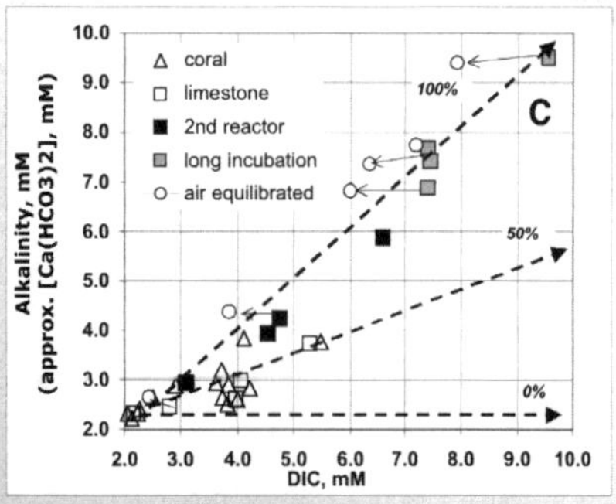

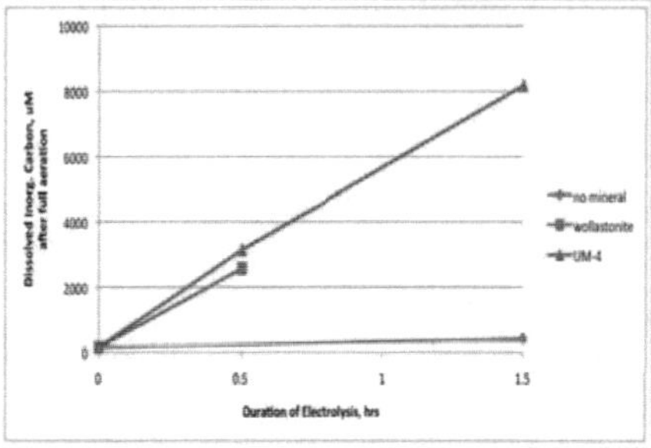

Uses of HCO_3^-

- ❑Environmental applications e.g. place in ocean to offset effects of ocean acidity
- ❑Use as substrate for algae biosynthesis of food/pharma/fuels
- ❑Use as substrate for chemical synthesis of hydrocarbons/fuels

Advantages over Straight CO_2 Addition

- ❑ Greater dissolved inorganic carbon concentration per input of CO_2 (left panel)
- ❑ Less pH depression per unit input of CO_2 (right panel)

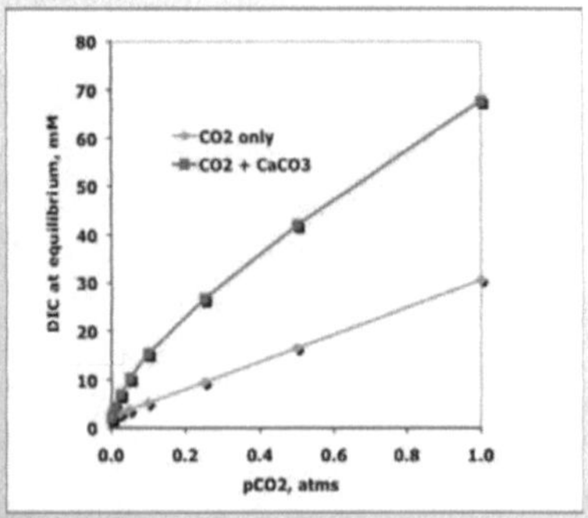

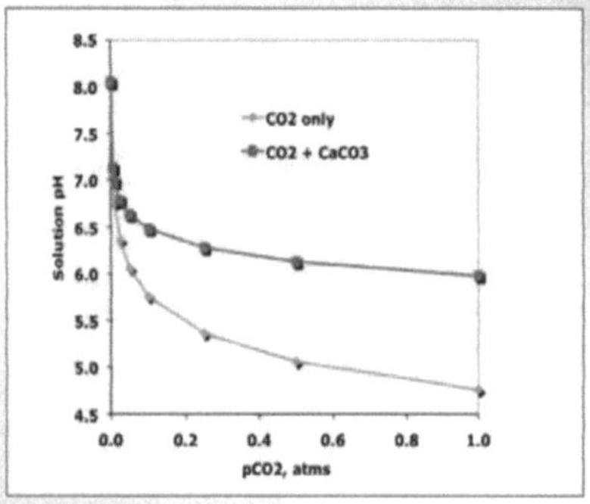

Equilibrium conditions modeled for seawater.

Overcoming C limitation in Algae: Experiments at LBL

- Aeration of lab scale culture with air or enriched CO_2 in the presence or absence of added $Ca(HCO_3)_{2aq}$

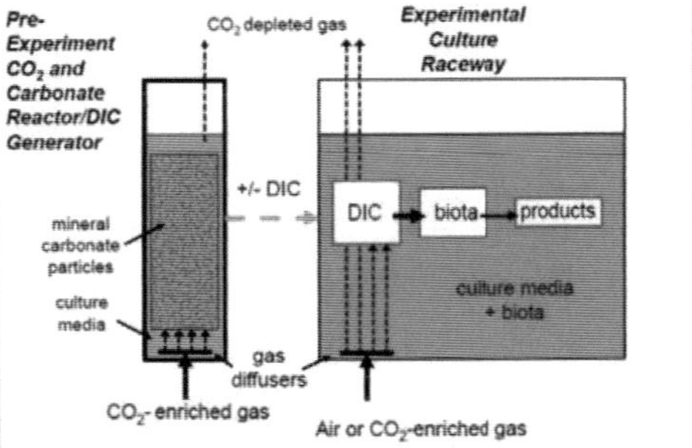

- Do algae grow faster/denser in the presencc of $Ca(HCO_3)_{2aq}$? Optimum concentrations and procedures?
- Can use of $Ca(HCO_3)_{2aq}$ allow adequate growth on air alone, i.e. avoid CO_2 derived from fossil fuel combustion?

Relevance to abiotic hydrocarbon fuel production

- **E.g., C limitation in artificial photosynthesis:**

"The only real sticking point in the entire artificial photosynthesis game is that artificial catalysts are still not able to efficiently utilize carbon dioxide in the concentrations that are currently in the atmosphere." Ken Sleight, reporting on this paper:

NANO LETTERS — Letter — pubs.acs.org/NanoLett

A Fully Integrated Nanosystem of Semiconductor Nanowires for Direct Solar Water Splitting

Chong Liu,[†,§] Jinyao Tang,[†] Hao Ming Chen,[†] Bin Liu,[†] and Peidong Yang*[,†,‡,§]

[†]Department of Chemistry and [‡]Department of Materials Science and Engineering, University of California, Berkeley, California 94720, United States

[§]Materials Sciences Division, Lawrence Berkeley National Laboratory, Berkeley, California 94720, United States

Supporting Information

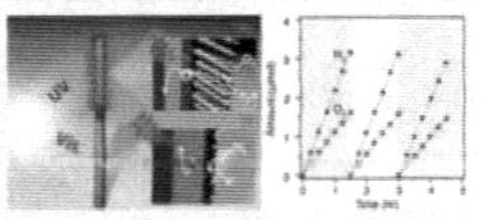

ABSTRACT: Artificial photosynthesis, the biomimetic approach to converting sunlight's energy directly into chemical fuels, aims to imitate nature by using an integrated system of nanostructures, each of which plays a specific role in the sunlight-to-fuel conversion process. Here we describe a fully integrated system of nanoscale photoelectrodes assembled from inorganic nanowires for direct solar water splitting. Similar to the photosynthetic system in a chloroplast, the artificial photosynthetic system comprises two semiconductor light absorbers with large surface area, an interfacial layer for charge transport, and spatially separated cocatalysts to facilitate the water reduction and oxidation. Under simulated sunlight, a 0.12% solar-to-fuel conversion efficiency is achieved, which is comparable to that of natural photosynthesis. The result demonstrates the possibility of integrating material components into a functional system that mimics the nanoscopic integration in chloroplasts. It also provides a conceptual blueprint of modular design that allows incorporation of newly discovered components for improved performance.

KEYWORDS: *Artificial photosynthesis, water splitting, nanowire-based heterostructure*

Could AWL technology provide a solution?

26

Conclusions

- Expensively making concentrated CO_2 from dilute sources and sequestering from air is not the only way to manage anthropogenic and atmospheric CO_2.
- Methods employing/adapting geo-chemistry have the proven capacity to significantly, safely, and cost-effectively contribute to reducing or consuming CO_2 emissions.
- Process engineering is needed to better determine optimum designs and operating procedures that maximize cost effectiveness, capacity, and environmental benefit.
- Partners and collaboration invited, there is a need for urgency - "the Earth is in the balance"

THANKS

UOP
A Honeywell Company

Process Development Symposium
May 1, 2014; Palo Alto, CA
AIChE NorCal Chapter

1914 - 2014 | A Century of Innovation

© 2014 UOP LLC. All rights reserved. UOP 6151-1

Agenda

UOP
A Honeywell Company

- **Introduction**
- **Technology Development Map**
- **Various Disciplines**
- **Putting pieces together**
- **Case Study**
- **Summary**

UOP LLC; A Honeywell Company

Primary Business: Develop processes and products for Natural Gas, Petroleum Refining and Petrochemicals

- **Established in 1914**
- **$3B Revenue in 2013**
- **Nearly 4,400 Employees Worldwide**
- **Core Capabilities**
  - Process Technology for Petroleum Refining, Petrochemical and Gas Processing industries
  - Catalyst and adsorbent products for petrochemical and petroleum refiners
  - Molecular sieves and alumina products for the manufacturing, chemical processing and natural gas industries
  - Research & Development
  - Engineering Consulting Services

Innovative Technologies for Sustainable Growth & Customer Satisfaction 1914-2014 A Century of Innovation

From Concept to Commercialization....

Development of a technology is an art of weaving chemistry, catalysis, unit operations, reactor Kinetics, modeling, process design, safety, Hazard analysis, demonstration, construction and Start-up -Often a Tortures Path

Establishing of an idea of technology need

uop
A Honeywell Company

Idea Generation

First Step

Technology Development Map

Ideas Come from Many Sources

Necessity is mother of invention

UOP 4050-5

Technology Development Map

UOP
A Honeywell Company

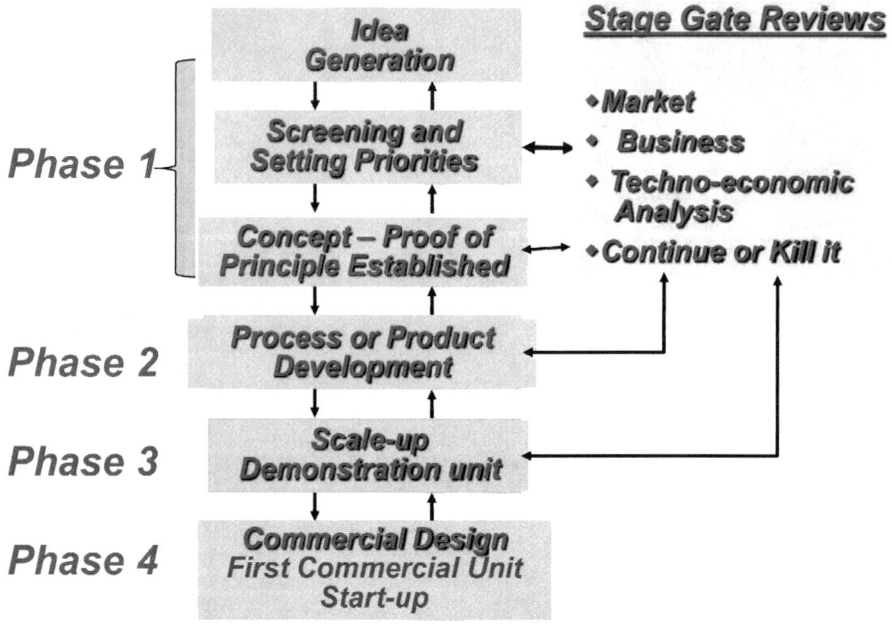

UOP 4050-26

Technology Concept to Commercialization

UOP
A Honeywell Company

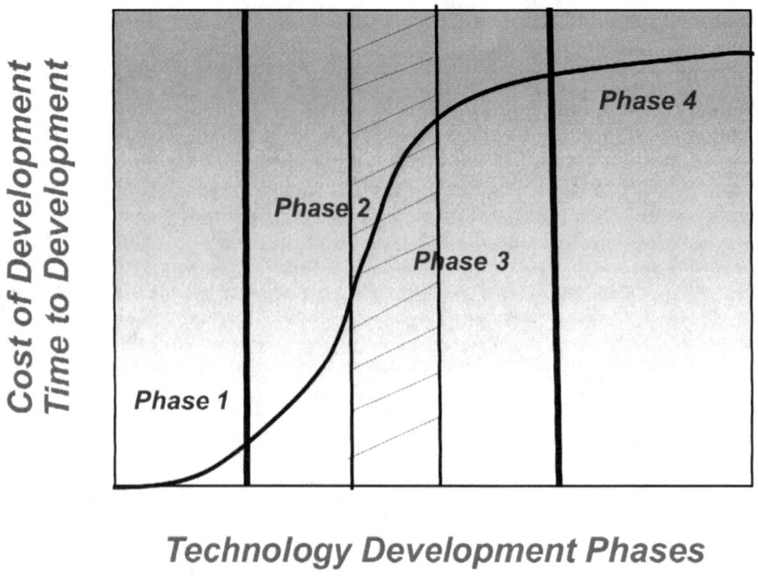

uop
A Honeywell Company

Technology Development Map

Do the Right Thing
Project Selection Is Very Critical

Otherwise it becomes a Costly and Time Consuming Error

Phase-1:
Proof of Principles - Chemistry & Catalysis

uop
A Honeywell Company

Metal and Acid Catalyzed Reactions by Sites in

Paraffin Dehydrogenation with - Unmodified Catalyst

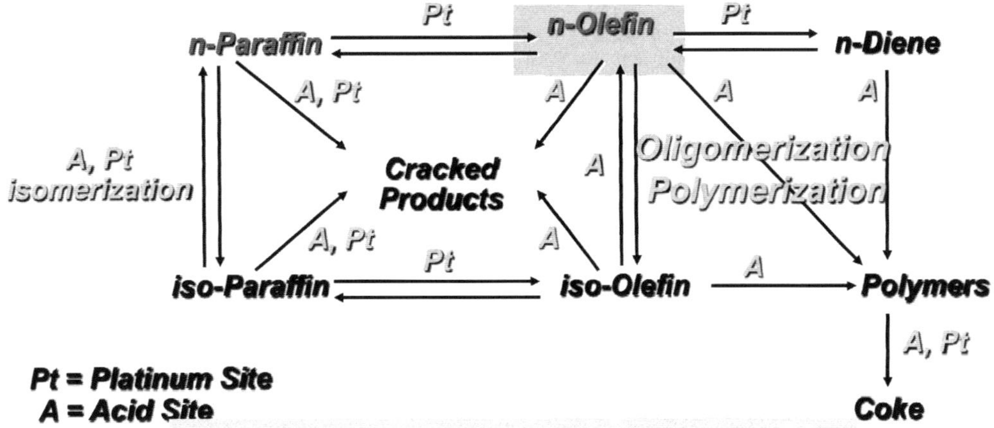

Pt = Platinum Site
A = Acid Site

Invention of a catalyst and defining operating conditions is not a simple task

UOP 4050-13

Selection of Process Conditions

UOP
A Honeywell Company

Propane Dehydrogenation Equilibrium

at 1.00 Atm Abs Pressure

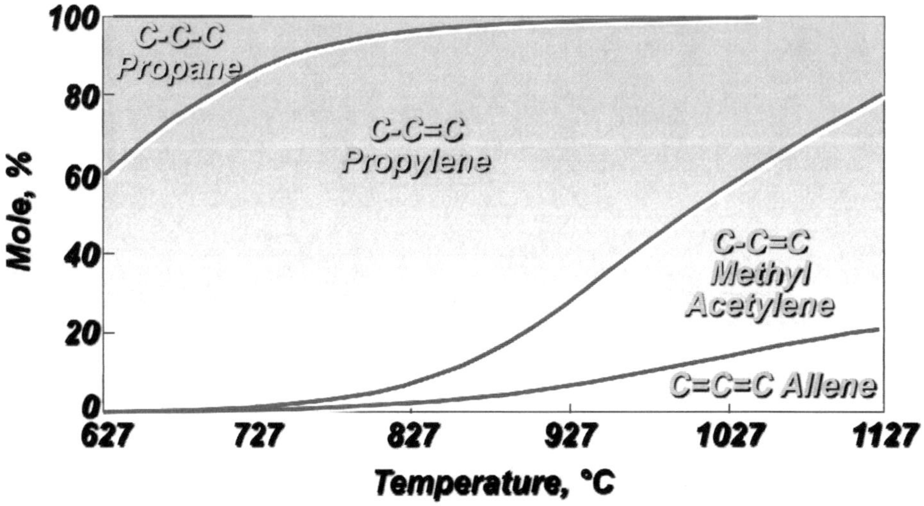

UOP 4050-15

Chemistry & Catalysis

UOP
A Honeywell Company

Optimization of Paraffin Dehydrogenation on Modified Pt Catalyst

Light Paraffin Dehydrogenation

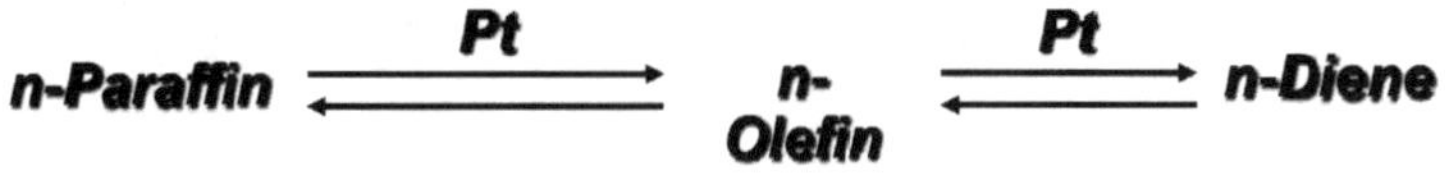

Heavy Paraffin Dehydrogenation

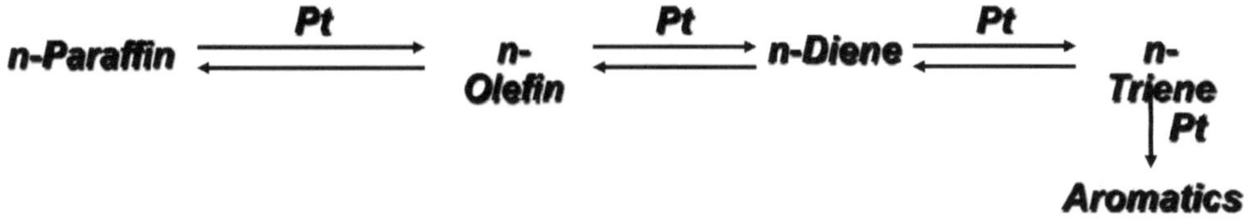

UOP 4050-12

Technology Development Map

UOP
A Honeywell Company

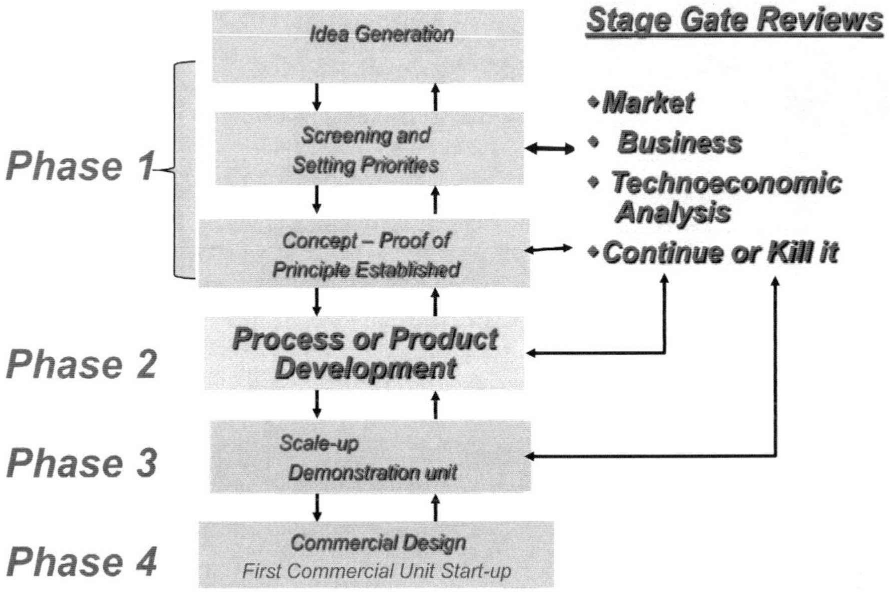

UOP 4050-26

A Complete Pilot Plant For Simulation Of Commercial Operation

UOP
A Honeywell Company

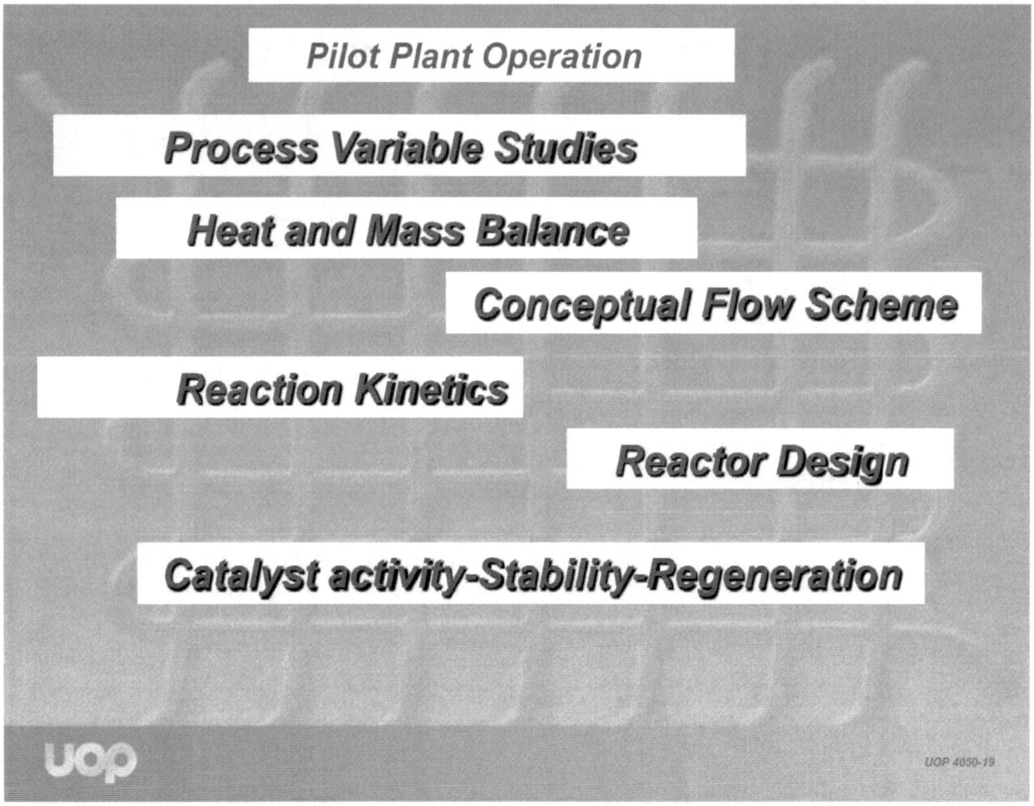

Reactor Design Options

uop
A Honeywell Company

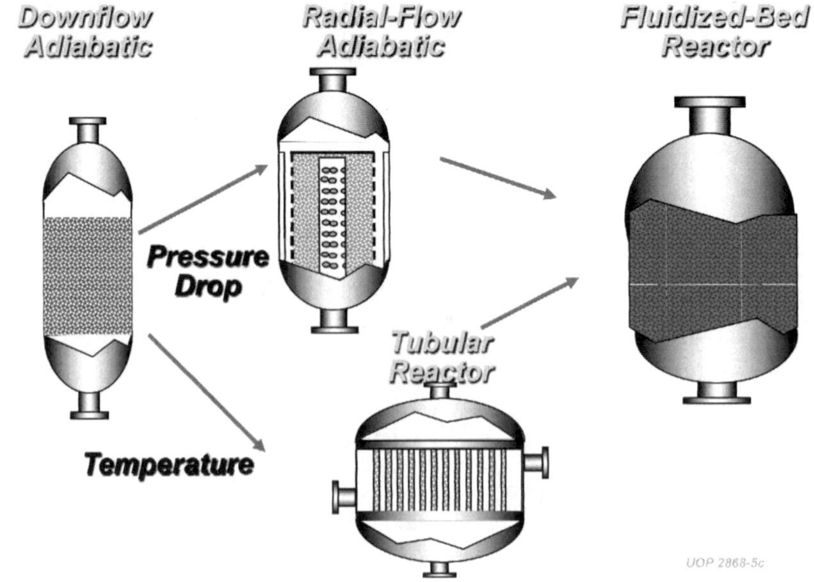

UOP 4050-22

Identify trace impurities- May need Analytical Development

uop
A Honeywell Company

Develop method for the analysis

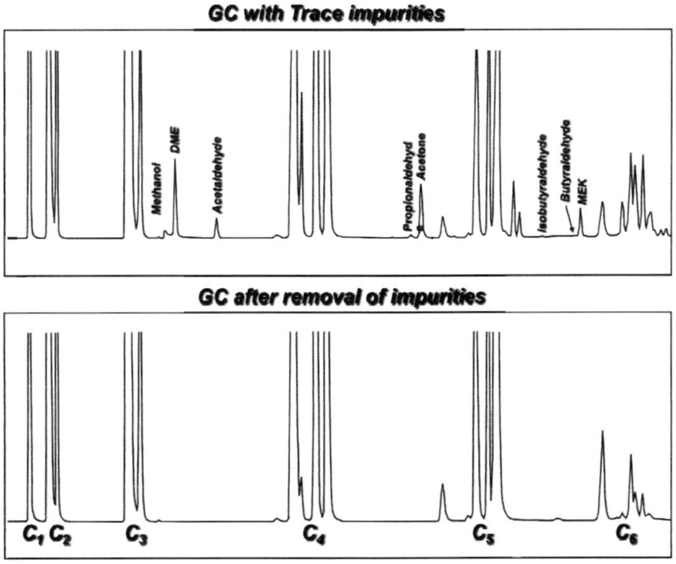

Identify & Quantify

Identify or invent a method for the removal

Technology Development Map

uop
A Honeywell Company

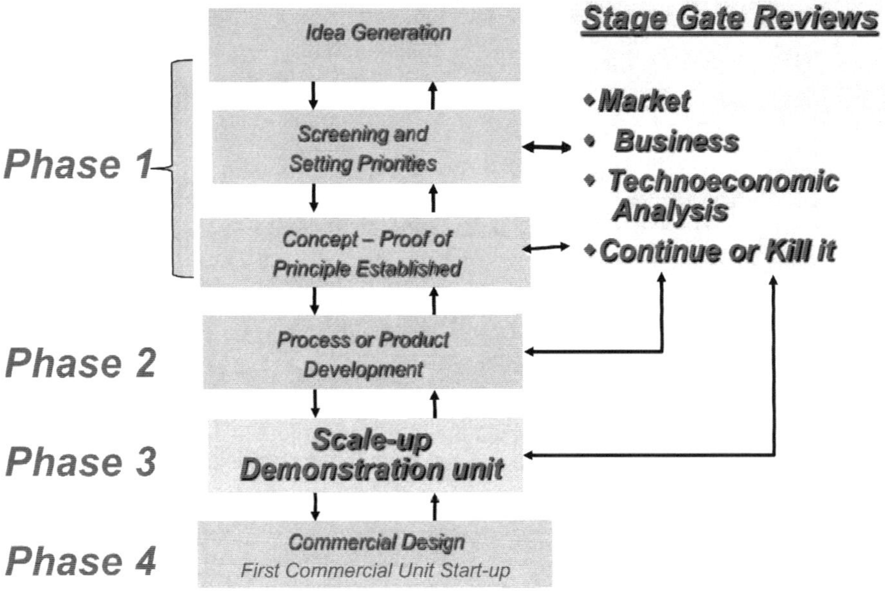

UOP 4050-26

From Concept to Commercialization

UOP
A Honeywell Company

Idea Generation *Proof of Principle Experiments,*
Technoeconomic analysis *Project Selection*
Process Development *Scale Up* *Detail Design*
Selection of Metallurgy

Chemistry *Catalysis* *Reactor Kinetics*
Process Variable Studies *Reactor design*
Fluid Mechanics – Computational Fluid Dynamics
Heat and Mass balances *Detail Mechanical Design*
Hazard Analysis *Demonstration Unit*
Operating Manual
Commercialization
Pre-start up Check out *Start Up*
Confirm Product quality and Plant capacity
Art of Weaving all of the above

UOP 4050-29

Management Responsibilities

UOP
A Honeywell Company

- Provide vision, foresight, and resources
- Establish goals
- Set up stage-gate reviews
- Establish stimulating environment
- Reward failures
- **I have not failed 700 times, I have not failed once. I have succeeded in proving that those 700 ways will not work. When I have eliminated the ways that will not work, I will find the way that will work.**
  Thomas Edison

UOP 4050-28

Project Management

UOP A Honeywell Company

- **Team Leader**
  - Trained in team functions and leadership
  - Champion of Project
  - Jack of all trades
  - Tracks and reports monthly progress to sponsor
- **Team members specialized in their function and trained in team-building concepts**

UOP 4050-30

UOP Integrated LAB Complex

UOP A Honeywell Company

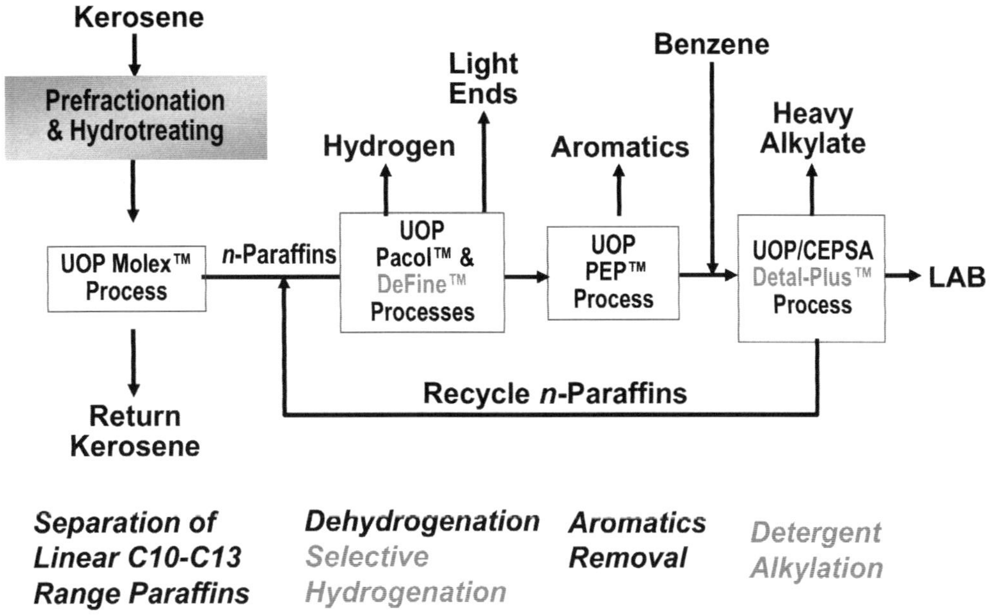

From Concept to Process: Incremental Development

UOP DeFine Process in LAB Production

UOP
A Honeywell Company

- *Improve LAB yield by selective hydrogenation of diolefins in the feed to alkylation*
- *Selective hydrogenation of acetylene and diolefins in Olefins industry well practiced*

- *With prior back ground in C2, C3, C4 acetylenes and diolefins hydrogenation, catalyst selection and operating parameters established in one month*

- *Prepare small samples of LAB with and without DeFine and establish yield and quality advantage*
- ***Contact first customer***

- *First commercial unit licensed*
- ***Construction***
- ***Start-up***

←——*One Month*——→ ←*One Month*→ *-22 months-*

Phase-1
Need identified Proof of Principle

Phase 2
Catalyst invention Lab scale and integrated pilot plant

Phase 3
Large demonstration unit and product qualification

Phase 4
Commercial design Construction and start-up of the first unit

UOP 6479-23

From Concept to Commercialization: UOP/Cepsa Detal Process

UOP
A Honeywell Company

- *1988 Motor Fuels HF Alkylation accident in a refinery*
- *HF safety highlighted*
- *Need to invent alternate catalyst*

- *Catalysis Research*
- *Lab Experiments*
- *Integrated pilot plant*
- *By 1990 Need for demonstration identified*

- *Cepsa of Spain joins the development and moves forward with a demonstration unit at their site in Spain*
- *1991-92 successful demonstration*
- *LAB product qualified by Detergent formulators*

- *1992 commercial design of first unit*
- *1994 start-up of first commercial unit in Canada*

---------------------------------*7 Years*---------------------------------

Phase 1
Need identified Proof of Principle

Phase 2
Catalyst invention Lab scale and integrated pilot plant

Phase 3
Large demonstration unit and product qualification

Phase 4
Commercial design Construction and start-up of the first unit

UOP 6479-24

Oleflex™ Process for Propane Dehydrogenation

UOP A Honeywell Company

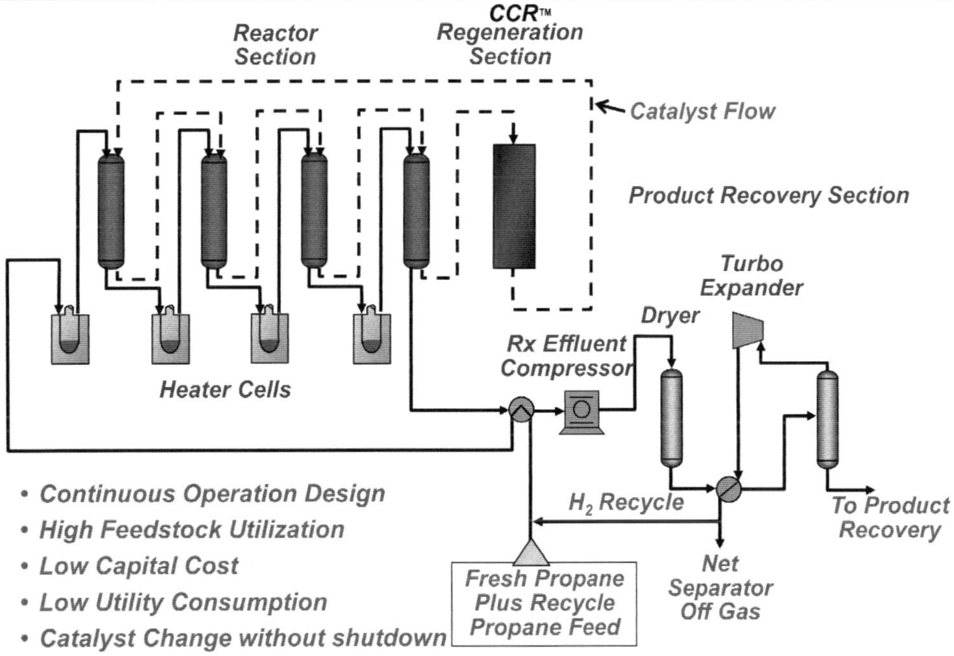

- *Continuous Operation Design*
- *High Feedstock Utilization*
- *Low Capital Cost*
- *Low Utility Consumption*
- *Catalyst Change without shutdown*

UOP - CONFIDENTIAL

From Concept to Process UOP Oleflex[R] Process for C3/iC4 Dehydrogenation

UOP A Honeywell Company

- *1973 Customer with significant propane wants to upgrade propane to propylene*
- *Approaches UOP*
- *R&D Project Established*

- *By 1976 significant PP work done*
- *Design of a process with Fluid bed & Cr catalyst established*
- *1977 Hexavelant Cr identified as carcinogen*
- *Project Redirected to inventing Pt catalyst*
- *By 1980 Pt catalyst and reactor with CCR identified*

- *1980-83 detail process design including mechanical design of Rx-regn, Kinetic modeling, Hazard analysis Completed*
- *No need of large demonstration as the technology similar to prior Pacol[R] and CCR-Platforming[R] process*

- *1987 First unit licensed*
- *1990 start-up of the first unit*
- *Trouble shooting*
- ***1990-1994 several more propane and isobutane dehydrogenation unit start-ups.***

----------------4 Years-----

--------------------13 years--------------------

Phase-1 Need identified Proof of Principle	Phase 2 Catalyst invention Lab scale and integrated pilot plant	Bypassed - Phase 3 demonstration unit	Phase 4 Commercial design Construction and start-up of the first unit

UOP 6479-26

UOP FCC Expertise employed in MTO

UOP
A Honeywell Company

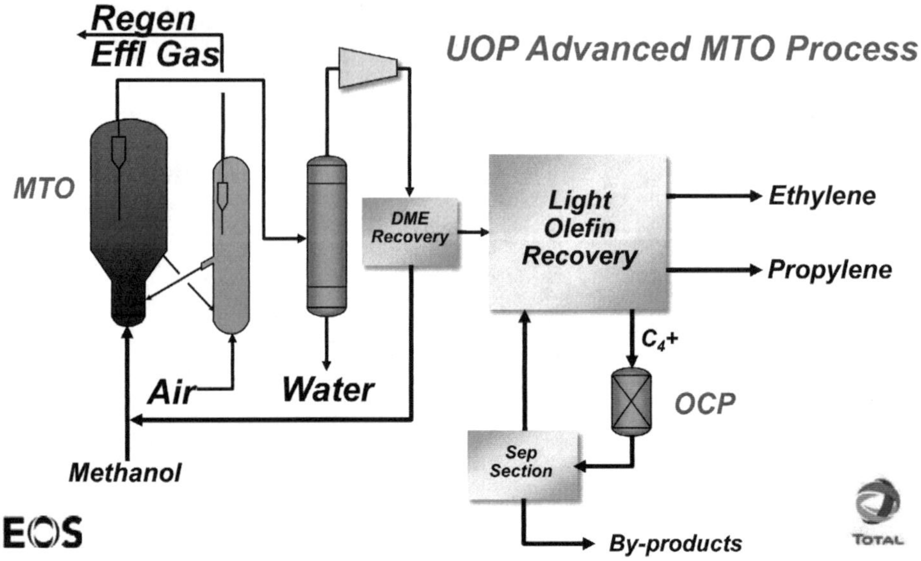

INEOS

TOTAL

UOP has licensed 4 MTO Units with the first to S/U in 2013

UOP 5970C-27

From Concept to Process UOP Advanced MTO

UOP
A Honeywell Company

- *~1 kg/day of MeOH feed*
- *Multiple grades/sources of feed*
- *Reactor and regenerator configuration*

- *~1 ton/day (maximum) of MeOH feed*
- *Multiple grades/ sources of feed*
- *Reactor and regenerator configuration*

- *~10 t/d of MeOH feed*
- *Fully integrated Advanced MTO Process Configuration*

- *2,300 t/d of MeOH feed*
- *Advanced MTO / OCP Configuration*

SAPO-34 Invented — 1983

UOP Pilot Plant Operation — *1988*

INEOS Demo Unit (at the time Hydro) — *1995*

Total Process Demo Unit — *2009*

Wison CEC Industrial Plant — *2013*

UOP 6479-28

Wisson Nanjing Advanced MTO Plant – October 2013

UOP
A Honeywell Company

UOP 6151-20

History of Propane Dehydrogenation

UOP
A Honeywell Company

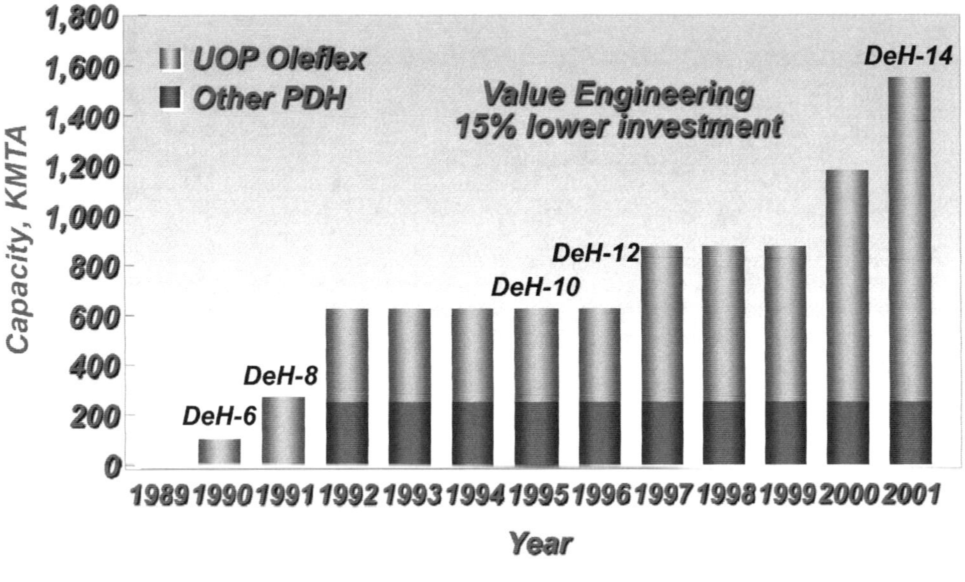

Continuous Innovations to stay competitive

LAB Technology
Story of Continuous Innovations

UOP
A Honeywell Company

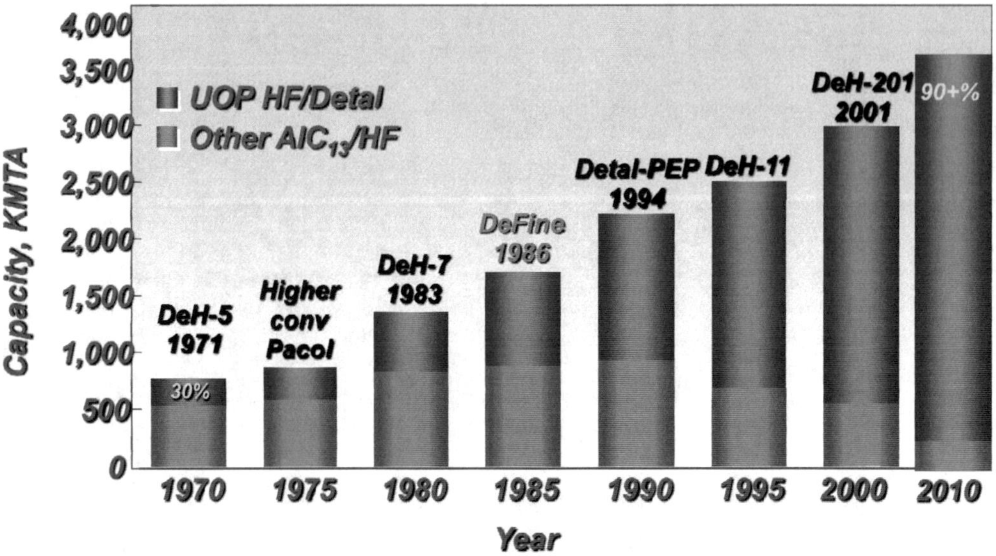

Continuous Innovation to move ahead of competition

UOP 4050-42

Time to Commercialization a New Technology

UOP
A Honeywell Company

- *Idea generation to Proof of Principle: 1-10 Yrs, If clear need and market potential identified,*
  it could be 1-3 Years
- *Process Development 1-5 Years*
- *Demonstration unit 3-10 Years*
- *Commercialization 3-10 years*

Total 5 To 25 + Years

Technology Commercialization Complexity Index

UOP
A Honeywell Company

Stage	*Relative Complexity Index*	*Time to Commercialize Years*
Early stage of Industry New Product-New process	Low	1-5
Environmental need or regulation promoted	low	1-5
Incremental improvements	Low	1-5
Govt. Mandate State ownership	Low	1-5
Growth stage- Reasonable Demand Growth	Medium	3-10
Mature Industry	High	5 to 25

Out of Box Thinking

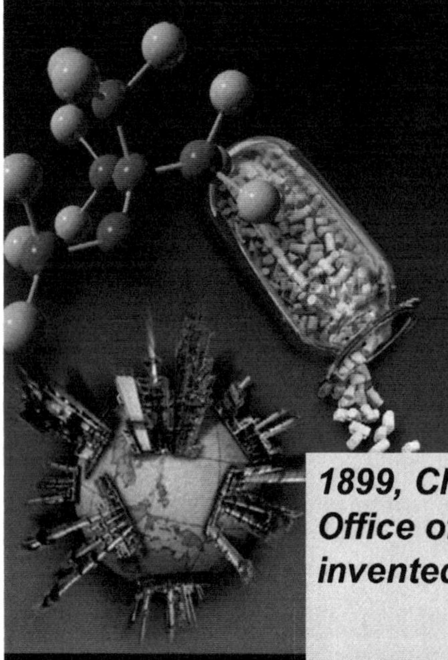

- Every generation has perceived limits to growth
- These limits are usually constrained by existing technology concepts
- We constantly fail to grasp how many ideas remain undiscovered and the impact of new technology on our lives

1899, Charles Duell, Commissioner, US Office of Patents. "Everything that can be invented has been invented"

uop

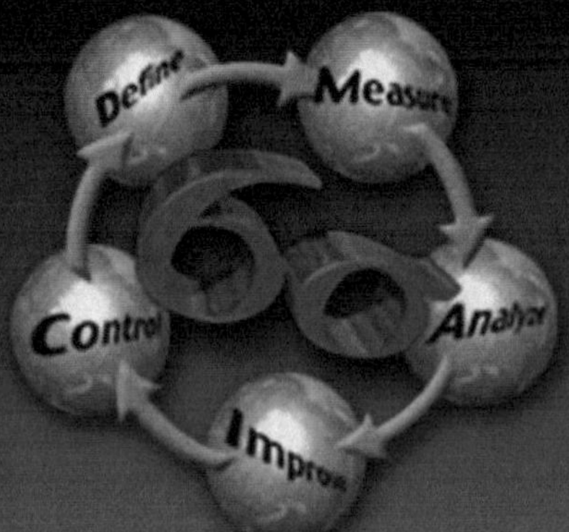

- Identify right projects to work on
- Employ modern tools
- Practice stage gate and six sigma methodology
- Have A champion follow it from concept to commercialization
- Reward failures, so success follows

uop

Commercialization of New Technology

UOP
A Honeywell Company

Research And Development Is A Serious Business

- *Not All Past Predictions Have Come True*
- *Visionary Management Is Key To Success*

Past Predictions

UOP
A Honeywell Company

1876 *"This 'telephone" has too many shortcomings to be seriously considered as a means of communication"* **Western Union Memo**

1895 *"Heavier-than-air flying machines are impossible"* **Lord Kelvin, President Royal Society**

1920 *"The wireless music box (radio) has no imaginable commercial value"* **David Sarnoffs Associates In Response To His Urgings For Investments In The Radio**

1943 *"I think there's a world market for maybe five computers"* **Thomas Watson, Chairman IBM**

1949 *"Computers in the future may weigh no more than 1.5 tons"* **Popular Mechanics Forecasting The Relentless March Of Science**

1977 *"There is no reason anyone would want a computer in their home"* **Ken Olson, President, Chairman And Founder Of Digital Equipment**

Acknowledgement

- Many thanks to my colleagues at UOP for reviewing this material
- My thanks to UOP for providing data, pilot plant and commercial unit photographs

Q&A

1914-2014 A Century of Innovation

UOP LLC, A Honeywell Company

UOP 6151-40

"It worked in the lab!", Case Studies in Commercialization

Presented to

American Institute of Chemical Engineers
Process Symposium May 1, 2014

by Ronald G. Bray

AIChE
The Global Home of Chemical Engineers

Agenda

- ***...."Where shall I begin, please your Majesty?" he asked.***
"Begin at the beginning," the King said gravely, "and go on till you come to the end: then stop."
(Alice's Adventures in Wonderland)

Commercialization Agenda

- Background
- Guide to Technology Transfer Packages
- Case study
- Questions

Type of Process Scale-up?

- New process for existing product ?
- New product with new process?
- Low Value product?
- High Value product?
- Batch or continuous ?
- Grass-roots, retrofit or outsource?

Ideal Project Flow

AICHE Process Symposium,2014_R.Bray

5

Actual Project Flow

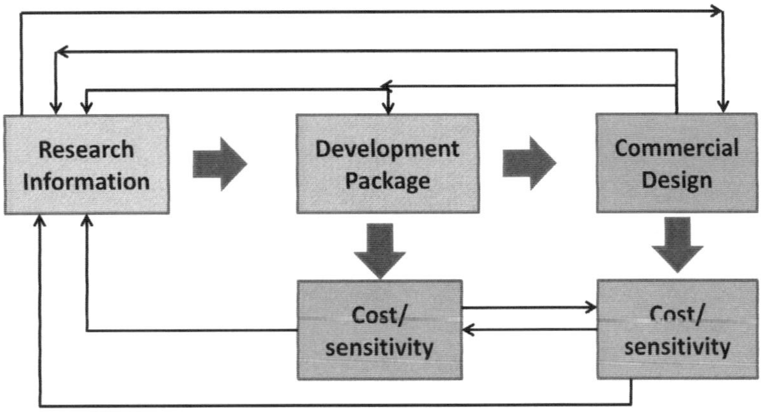

AICHE Process Symposium,2014_R.Bray

6

Research Technology Transfer Package(RTTP) Basic Information

- Chemistry of reaction(s)
- Laboratory procedure/Lab notebook
  - Reactants
  - Intermediates
  - Products
  - Temperature
  - Pressure
  - Yield
  - Waste
- Analytical Procedures
  - Sample preparation
  - Methods and equipment

AICHE Process Symposium,2014_R.Bray

7

Technology Transfer Package (TTP) Enhanced Information

- Physical properties
  - Reactants
  - Intermediates
  - Products
    - Molecular weight
    - Boiling point
    - Melting point
    - Solubility
  - Stability, time/temperature, pH
  - Side reactions
  - Wastes generated
  - Calorimetric properties (Exothermic, Endothermic)
  - Reaction characteristics-foaming, boiling, solids formation
  - Toxicology-MSDS
  - Safety
- Product specifications
- Reactant specifications

AICHE Process Symposium,2014_R.Bray

8

Optimization Mantra

- More economical starting materials
- Higher productivity
- Higher yield
- Shorter reaction time
- Higher concentration
- Less solvent
- Less waste

AICHE Process Symposium,2014_R.Bray

9

Engineering TTP

- Heat and material balance
  - Calorimetric Information
  - Simulation results
- Process description
- Process operating ranges
- Reaction kinetics
- Waste treatment
- Detailed Design
  - Equipment sizing
  - Control scheme
  - Piping and instrument Diagrams
  - Safety Analysis, (PHA, HAZOP)

AICHE Process Symposium,2014_R.Bray

10

What gets us into trouble is not what we don't know, it's what we know for sure that just ain't so?-
Mark Twain

Asparagine Production

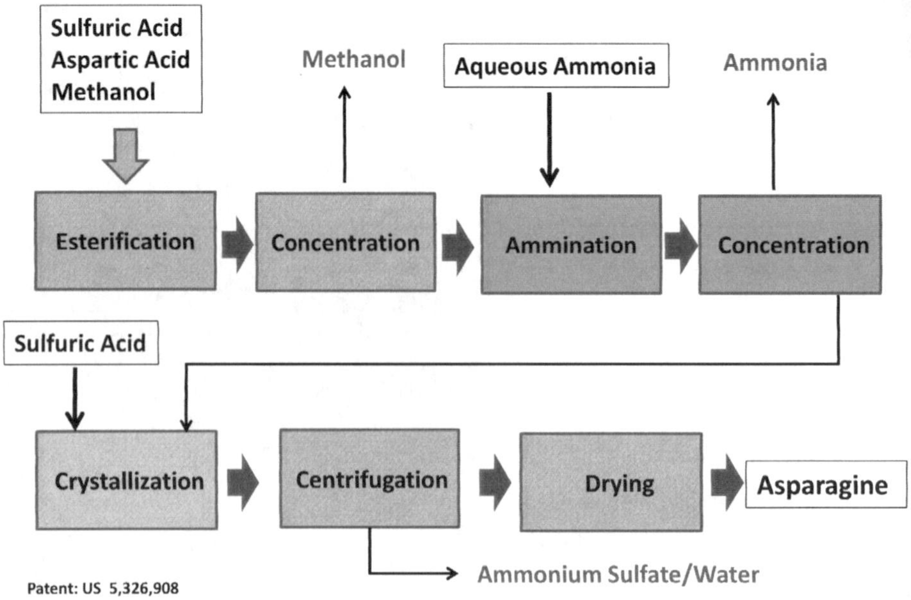

Patent: US 5,326,908

Laboratory Process

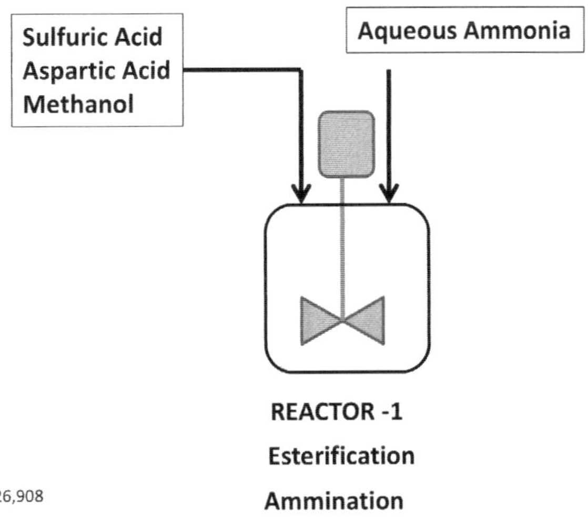

Patent 5,326,908

Pilot Plant Process

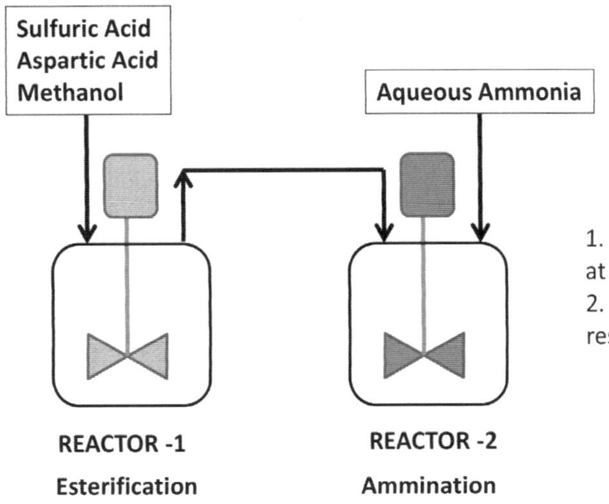

1. Process run 10X at 400 L.
2. Good repeatable results

Commercial Plant Process

It worked in the laboratory!

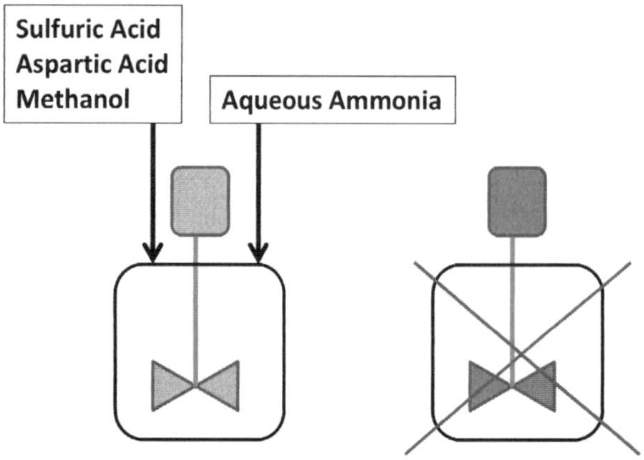

30,000 Liters/8000 Gallon Reactor

REACTOR -1

Esterification

Ammination

REACTOR -2

AICHE Process Symposium,2014_R.Bray

15

Many want credit for success, none want to be blamed for failure- *Anonymous*

- It worked in the laboratory!
- It's an engineering problem!

AICHE Process Symposium,2014_R.Bray

16

Respect your Reaction Kinetics!!-RGB

1. Asp-Ester-acid salt +NH_4OH→Asparagine-ammonia salt
2. H_2SO_4+ 2*NH_4OH→ $(NH_4)_2SO_4$ + H_2O
3. Asp-Ester-acid salt+ H_2O + NH_4OH→ Aspartic acid +$(NH_4)_2SO_4$ + H_2O

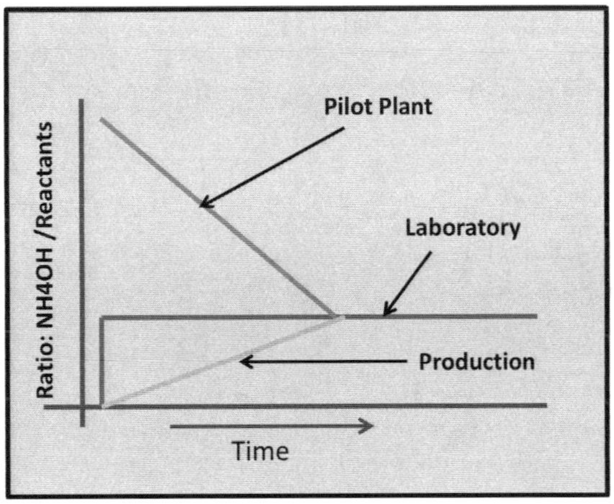

AICHE Process Symposium,2014_R.Bray

17

Revised Commercial Process

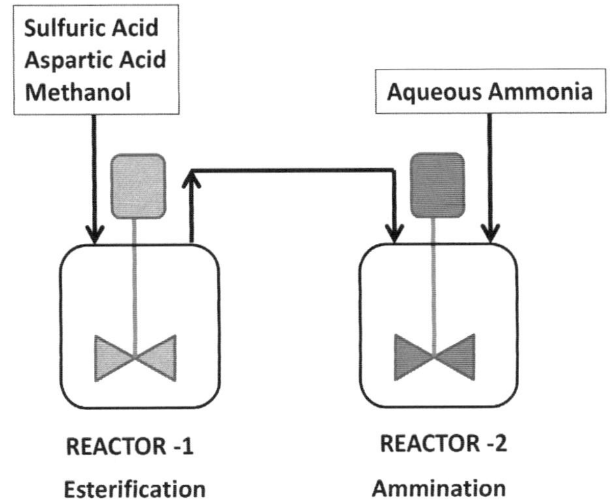

AICHE Process Symposium,2014_R.Bray

18

"Top phrases during scale-up"

- "Reduce to dryness."
- "Stir 16 hours (overnight)"
- It's just an engineering problem!
- I've never seen that before!
- I don't know!
- It's a little exothermic!
- It worked in the lab!

AICHE Process Symposium,2014_R.Bray

19

RAMO

- **R**epeatability
- **A**ccessibility
- **M**aintainability
- **O**perability

AICHE Process Symposium,2014_R.Bray

20

Thank you-
Questions?

AICHE Process Symposium,2014_R.Bray

21

The Semiconductor Industry from a Chemical Engineering Perspective

Dr. Zhihong Wang, Applied Materials

Originally presented on: 5/1/2014 12:30:00 - 12:55:00

In this talk, a couple of case studies are presented to show how chemical engineering principles can be applied to understand and solve critical issues in semiconductor industry. Two topics, microdefects formation in crystalline silicon and chemical mechanical polishing (CMP), will be covered with more focus on CMP. CMP is a critical and enabling technology in modern semiconductor fabrication. Brief introduction of CMP technology will be covered, then a couple of specific CMP topics will be discussed from a chemical engineering perspective.

Direct Ink Writing of Designer Materials

Presented by Cheng Zhu

AICHE NorCal Process Development Symposium
Work Processes and Smart Manufacturing Session

Lawrence Livermore National Laboratory

Chris Spadaccini, Josh Kuntz, Eric Duoss, Wayne King, Melissa Marggraff, Robert Maxwell, Tom Wilson, Todd Weisgraber, Andrew Pascall, Cheng Zhu, Jonathan Hopkins, Rayne Zheng, James Frank, Joshua DeOtte, Chris Harvey, Tom Metz, Kyle Lange, Marcus Worsley, George Farquar, Tammy Olson, Sergei Kucheyev, Chris Orme, Kyle Sullivan, William Smith, Maxim Shusteff, Luis Zepeda-Ruiz, Scott Fisher, Tom Wilson, Alex Gash, & John Vericella

Academic Partners: Professors J. Lewis (Harvard), N. Fang (MIT), D. Tortorelli (UIUC)

LLNL-PRES-653709

This work was performed under the auspices of the U.S. Department of Energy by Lawrence Livermore National Laboratory under contract DE-AC52-07NA27344. Lawrence Livermore National Security, LLC

The world is on the cusp of a manufacturing revolution

Factory Model	Assembly Line Model	Digital Customized Manufacturing

Manufacturing environment

1800 → 1900 → 2000 → Future

Workers coming to 'factories' to operate powerful machines that produced goods	Henry Ford introduces the assembly line concept for manufacturing & mass production Further enabled by advances in robotic systems	Numerous, dispersed, autonomous, flexible and potentially networked manufacturing centers Enabled by advanced Computer-Aided-Design, Modeling & Simulation and Flexible Manufacturing Technology

Confluence of HPC , advanced design methodologies, & new processes will accelerate this revolution resulting in rapidly fabrication of components with unprecedented performance.

3D Printing is the process of joining materials to make objects from 3D model data, usually layer upon layer

- 3-D graphical model, parts built in layers
- No tools, dies, or forms
- Near final shape
- Reduced delivery times
- Mechanical properties equivalent to wrought
- Reduced material use, reduced waste
- Reduced inventory
- Significant cost & energy savings

Lawrence Livermore National Laboratory

Stereolithography
3D Systems

Selective Laser Sintering:
3D Systems

Fused Deposition
Stratasys

PolyJet Process
Objet

Broad range of **commercial 3D printers** and solidification schemes (photocuring, ΔT, laser sintering, drying, etc.)

Powder Bed Printing:
Z Corp, ExOne

Laser Net Shaping
Optomec

Direct Laser Sintering :
EOS, Concept Laser

Electron Beam Melting
Arcam

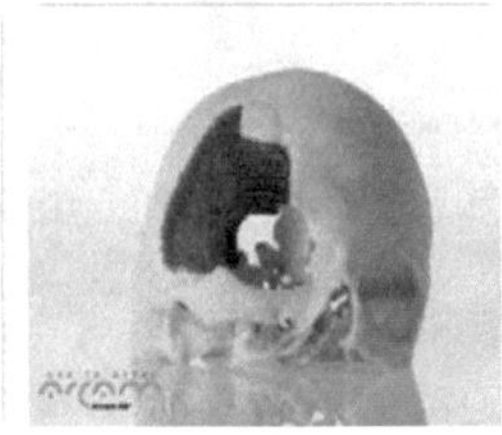

Lawrence Livermore National Laboratory

3D Printing is entering an exponential growth phase with low barrier to entry

Printed drones

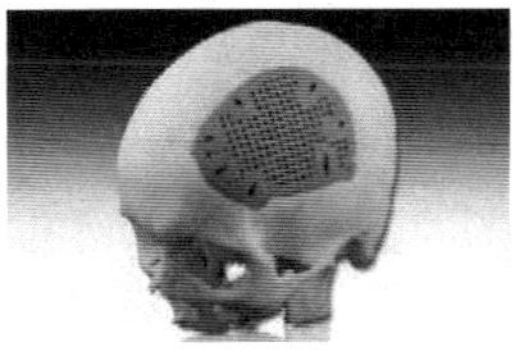

Medical implants

Optimized heat exchanger

- Polymer systems ~$1K - $300K; Metal systems ~ $1M
- Available globally

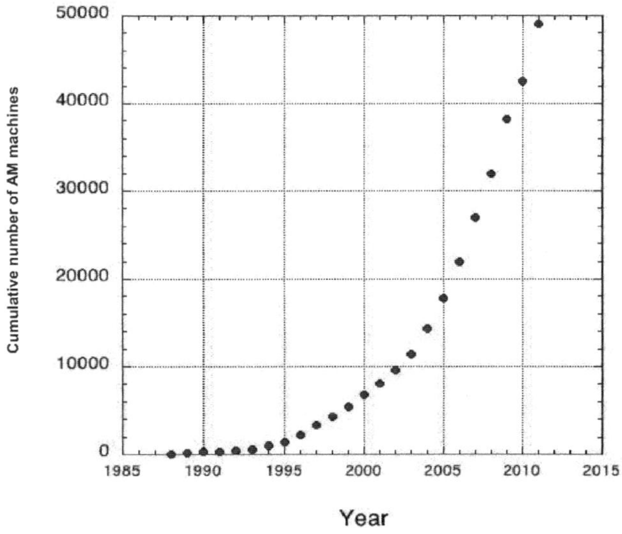

Lawrence Livermore National Laboratory

5

LLNL have developed custom additive micro- and nano-manufacturing technologies for engineered materials

Projection Microstereolithography (PμSL)
A photochemical and optical technique

Electrophoretic Deposition (EPD)
Electric fields transport nanoparticles

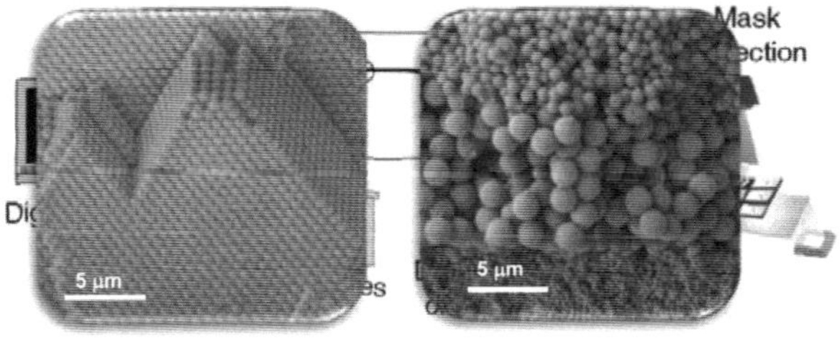

Direct Ink Writing (DIW)
Utilizes unique flow and gelling properties

Lawrence Livermore National Laboratory

6

Projection Micro-stereolithography (PμSL) - a photochemical and optical technique

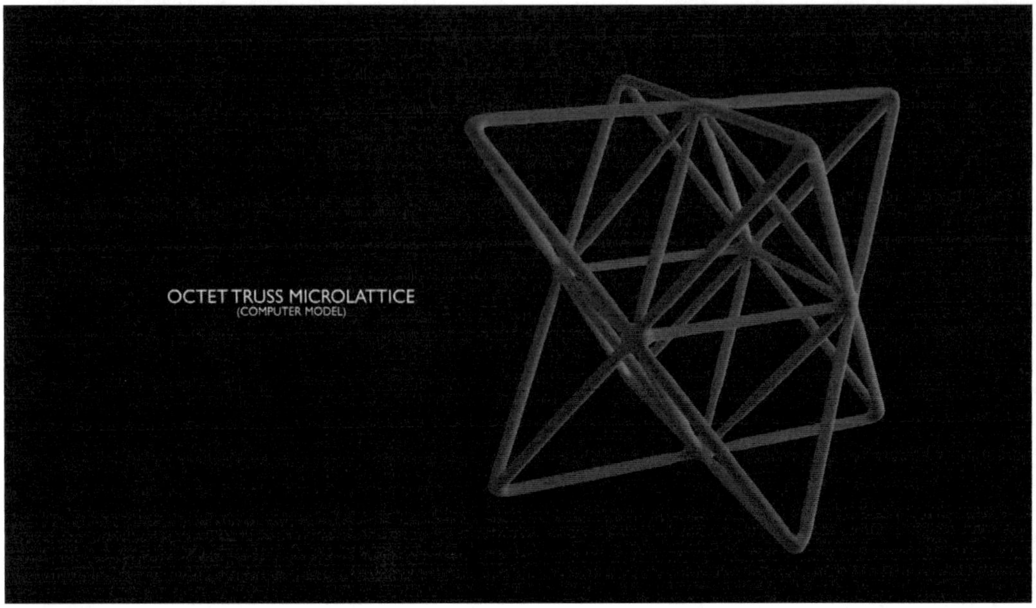

Lawrence Livermore
National Laboratory

7

Electrophoretic Deposition (EPD) - electric fields transport nanoparticles

Lawrence Livermore
National Laboratory

8

Direct Ink Writing (DIW) - utilizes unique flow and gelling properties

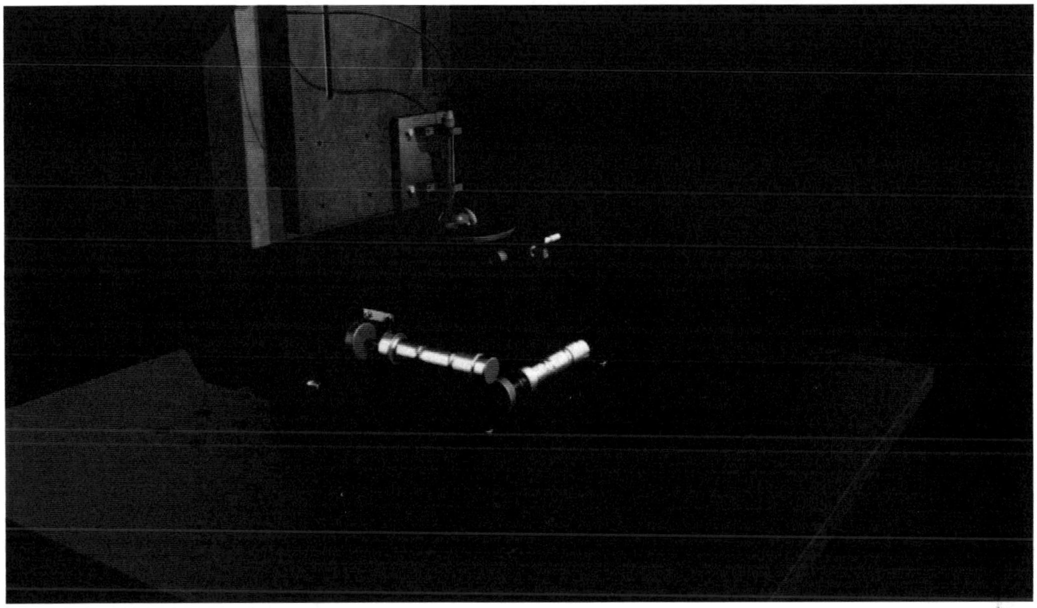

Lawrence Livermore National Laboratory

9

Ink properties required for DIW

Ink design and deposition:

- Ink must be able to flow through micro-nozzles without jamming;
- Ink filaments must form high integrity interfaces;
- Ink must solidify rapidly (via gelation, coagulation, or evaporation);
- Ink must has high concentration to minimize shrinkage during drying.

Desired Ink Rheology:

- Shear thinning behavior facilitates flow through fine nozzles without clogging
- Viscoelastic behavior enables printing of self-supporting (spanning) features

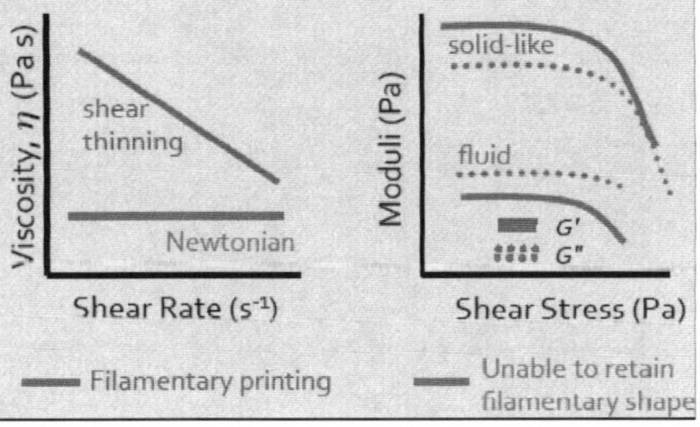

Lawrence Livermore National Laboratory

10

Energy absorbing materials

Defense/Aerospace | Sporting/Consumer Goods | Packaging/Transportation

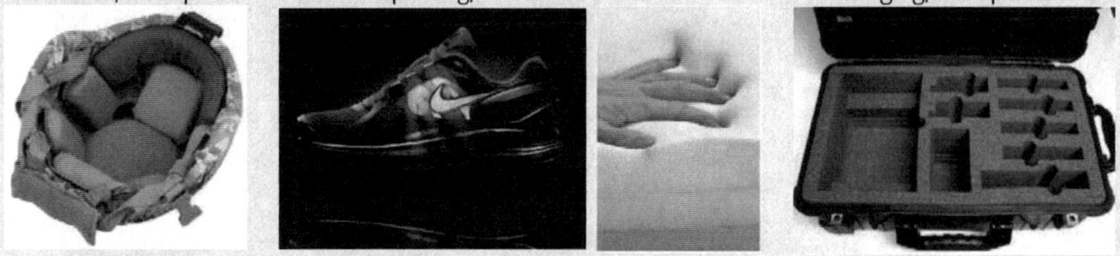

- Dampen shock and vibrations
- Distribute and relieve stress
- Maintain relative positioning
- Mitigate the effect of size variation

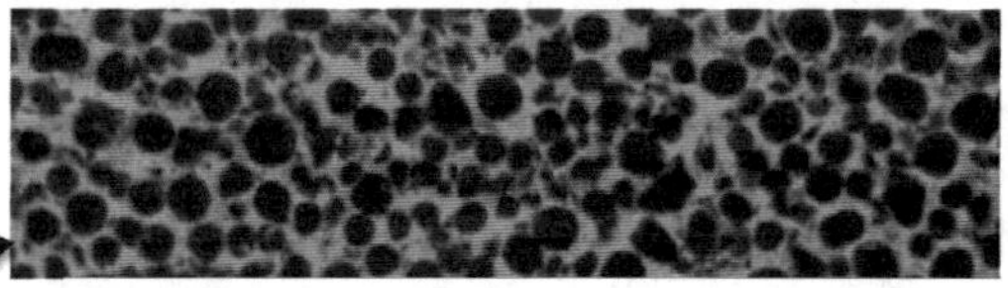

Foams are also ***stochastic***:

1. Difficult to control mechanical properties
2. Exhibit non-uniform properties
3. Little control over directional properties
4. Difficult to develop predictive models

Ordered cellular materials should:

1. Enable better control of mechanical properties
2. Exhibit more uniform properties
3. Enable better control over directional properties
4. Enable a predictive modeling capability

Lawrence Livermore National Laboratory

11

DIW of siloxane-based ink

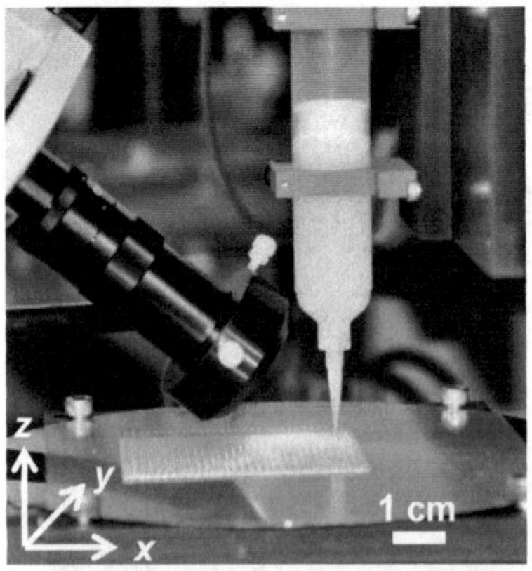

We targeted an ink that:

1. Would flow through micronozzles at reasonable pressures for long periods without clogging
2. Would rapidly solidify to maintain its shape
3. Would ultimately form an elastomeric material with high mechanical strength and elongation

We identified a commercially available material (Dow Corning SE1700) that meets these criteria.

DC SE1700 is a polydimethylsiloxane adhesive that is:

1. Two-part, silica-filled paste
2. Platinum catalyzed, addition cure (no byproducts)

Lawrence Livermore National Laboratory

12

Time-lapse video of printing process (610 μm nozzle, ~1 h build time)

Silica filler concentration is important for printability: Effect of diluting ink with unfilled siloxane resin

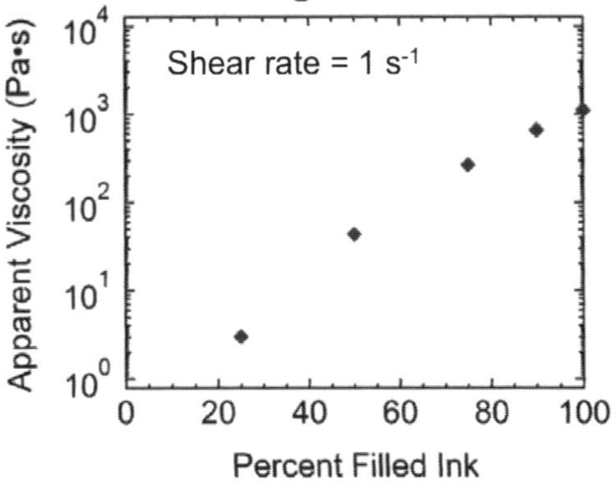

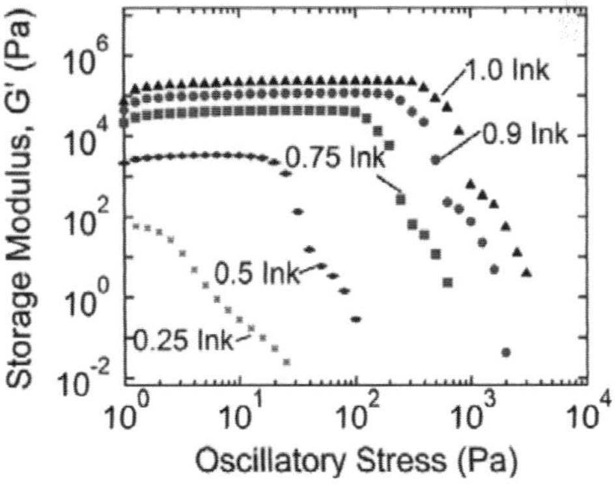

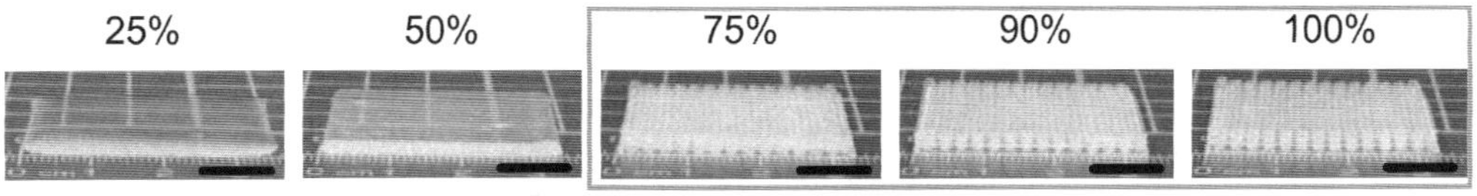

≥0.75 Ink is suitable for DIW

1.0 Ink will span ~2 cm gaps

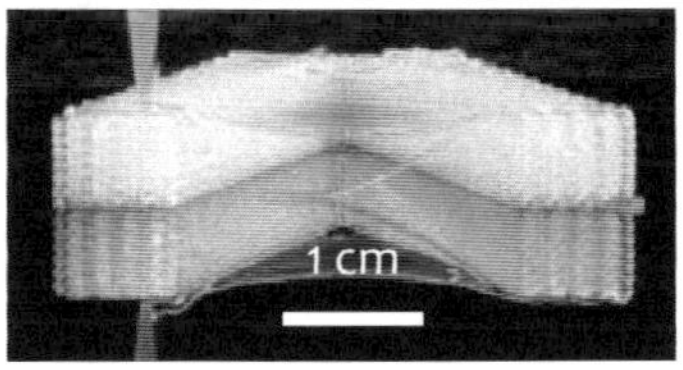

We patterned two micro-architectures and explored their mechanical properties

"Simple cubic" (SC) | "Face centered tetragonal" (FCT)

Uniaxial compression

Pre-compression + shear

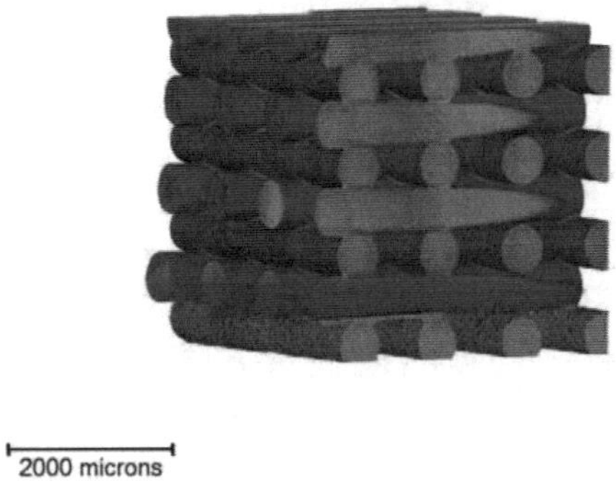

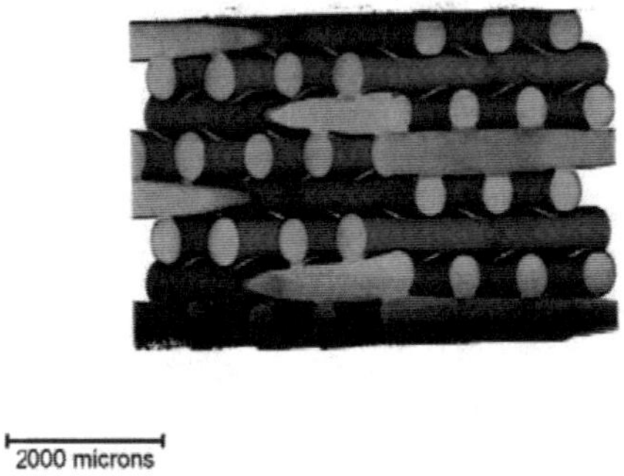

Lawrence Livermore National Laboratory

15

We patterned two micro-architectures and explored their mechanical properties

"Simple cubic" (SC) | "Face centered tetragonal" (FCT)

Uniaxial compression

Pre-compression + shear

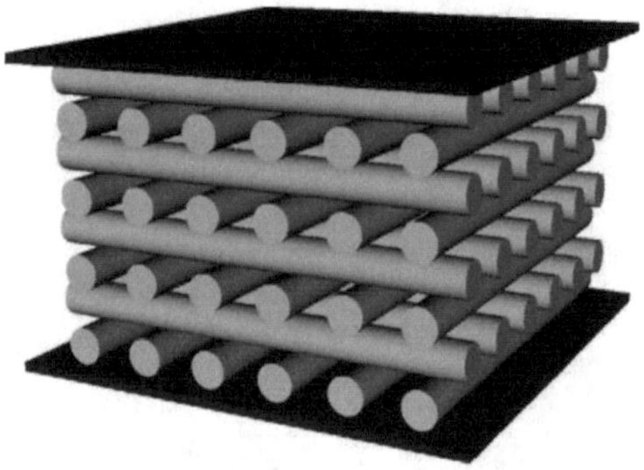

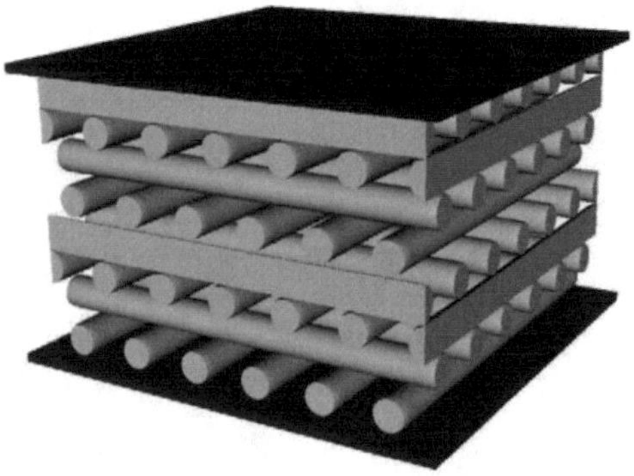

Lawrence Livermore National Laboratory

16

Architecture-dependent uniaxial compression behavior

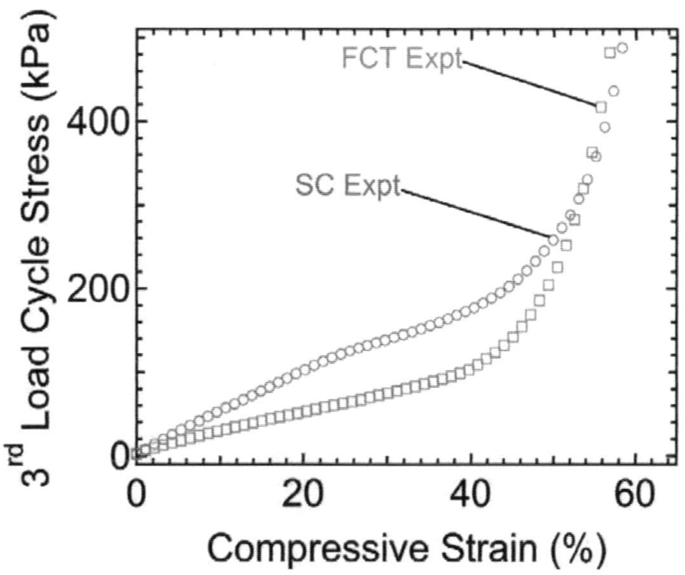

Lawrence Livermore National Laboratory

17

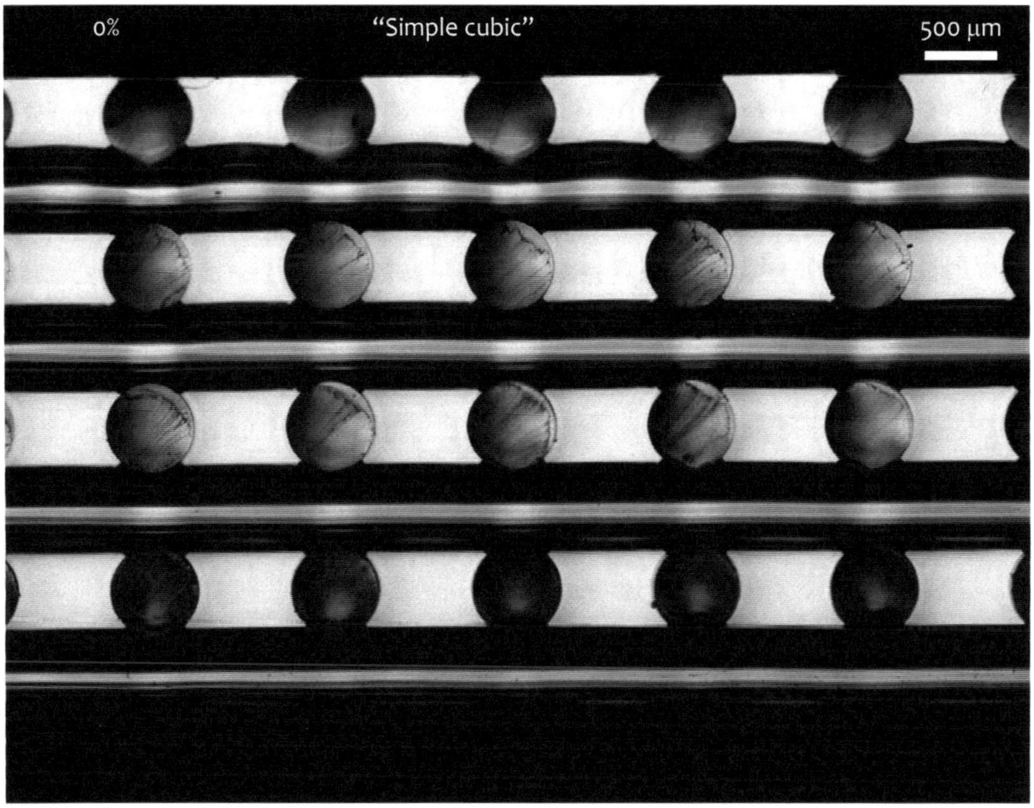

5%
"Simple cubic"
500 μm

10%
"Simple cubic"
500 μm

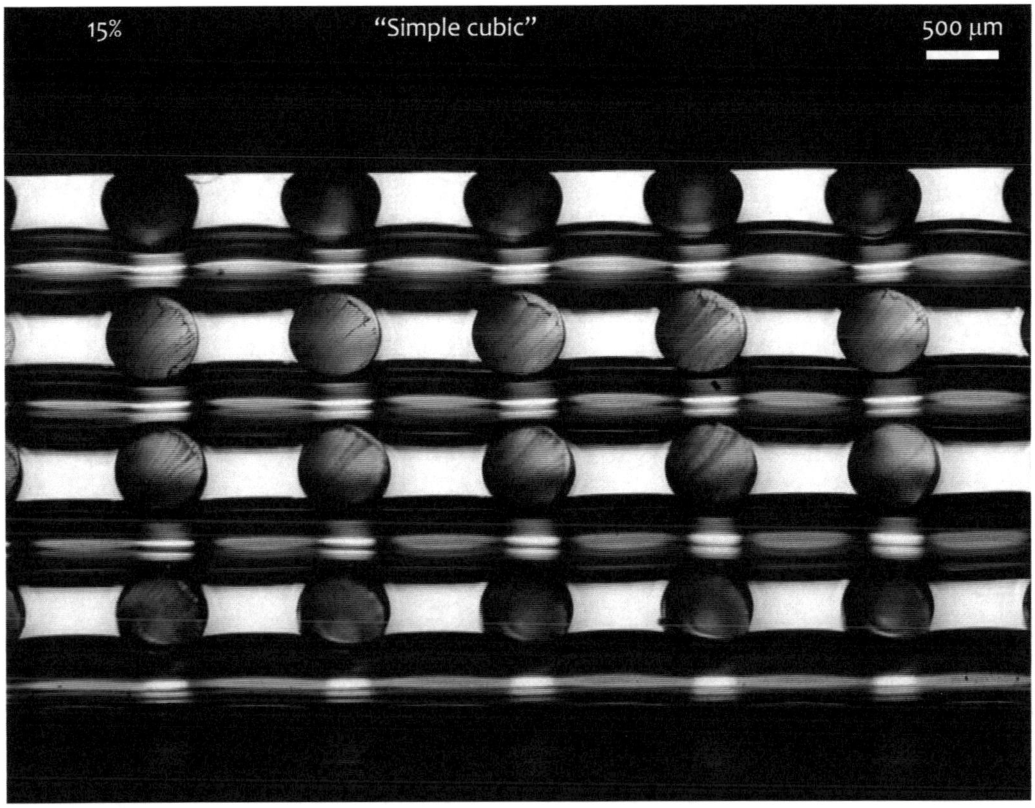
15%
"Simple cubic"
500 μm

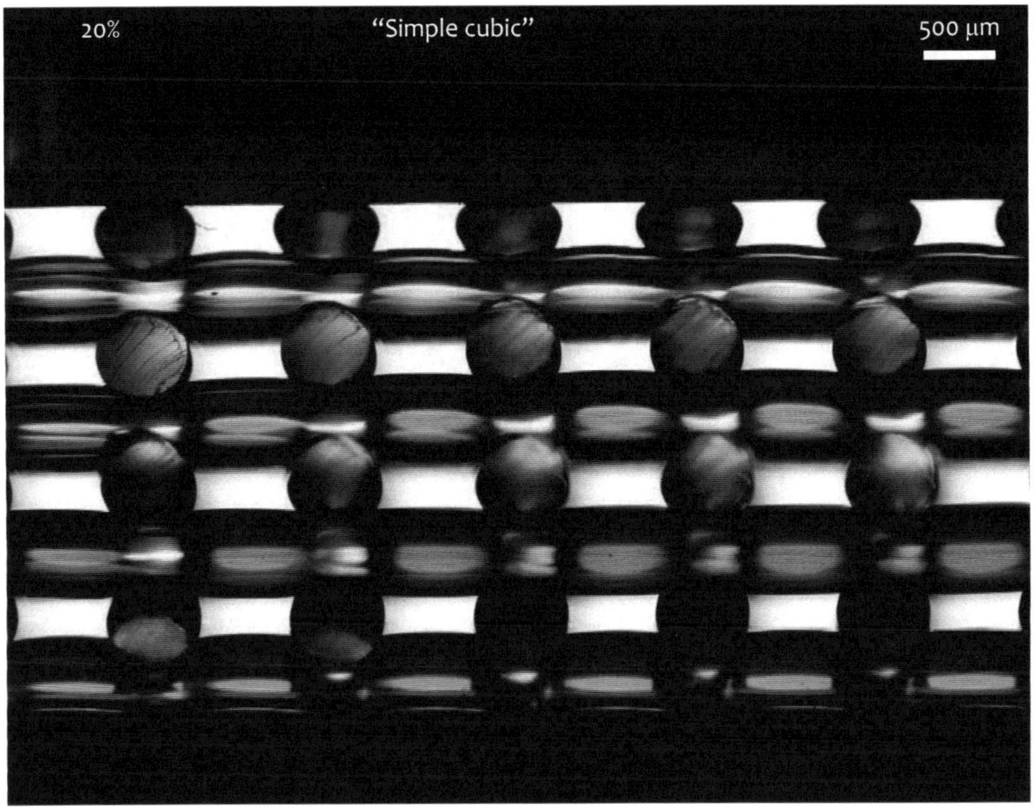
20%
"Simple cubic"
500 μm

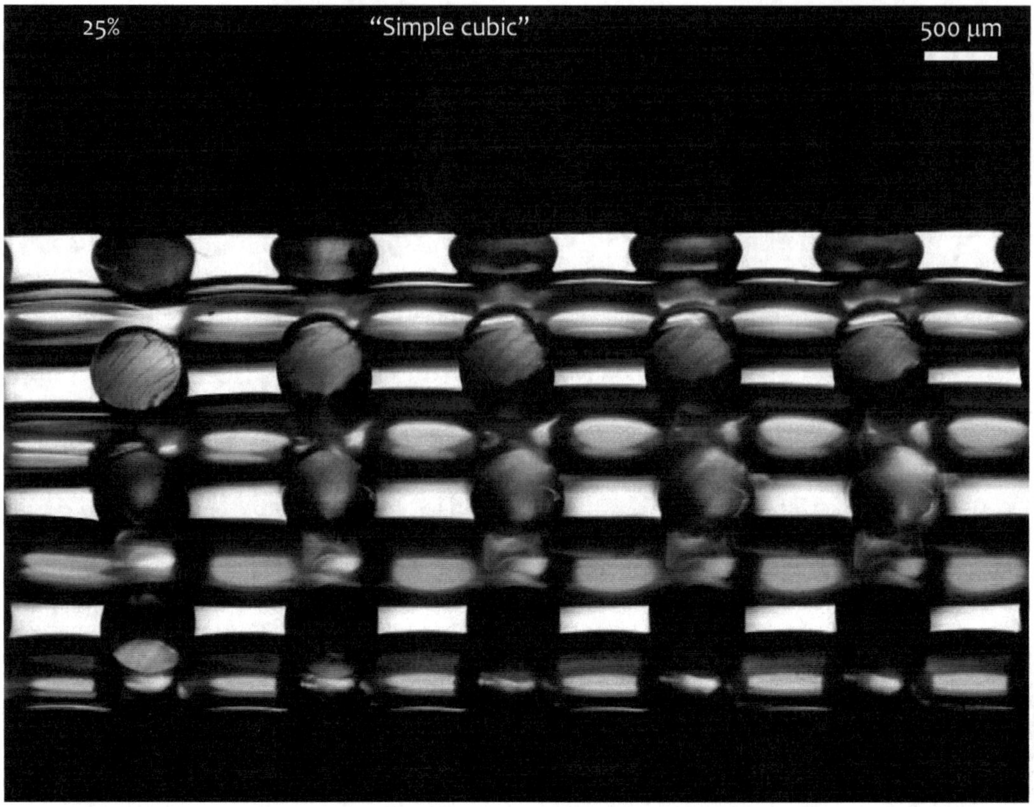

"Simple cubic"

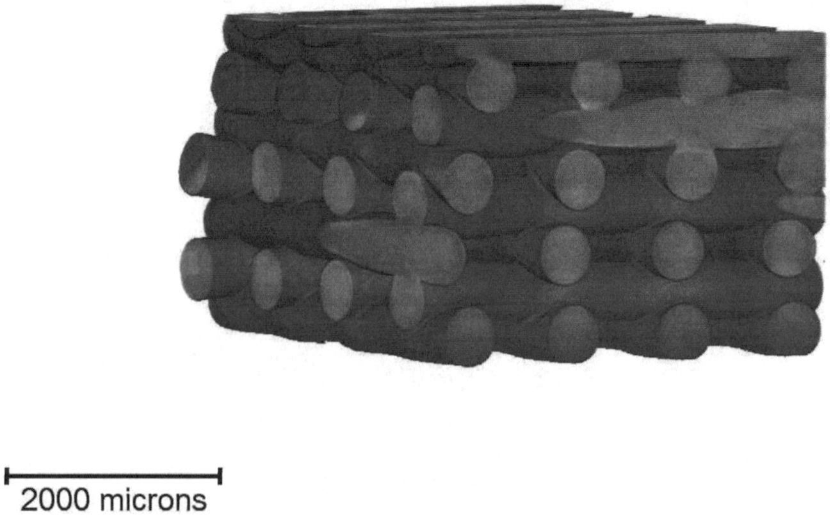

0%
"Face centered tetragonal"
500 μm

5%
"Face centered tetragonal"
500 μm

10%
"Face centered tetragonal"
500 µm

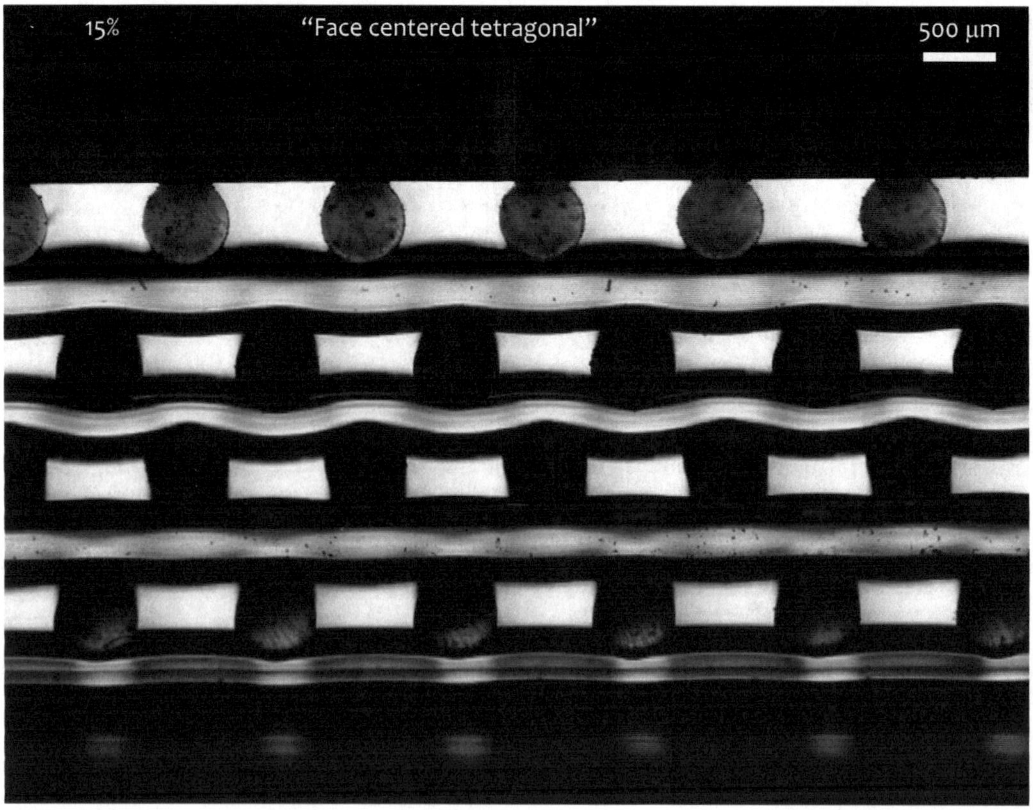
15%
"Face centered tetragonal"
500 µm

20%
"Face centered tetragonal"
500 µm

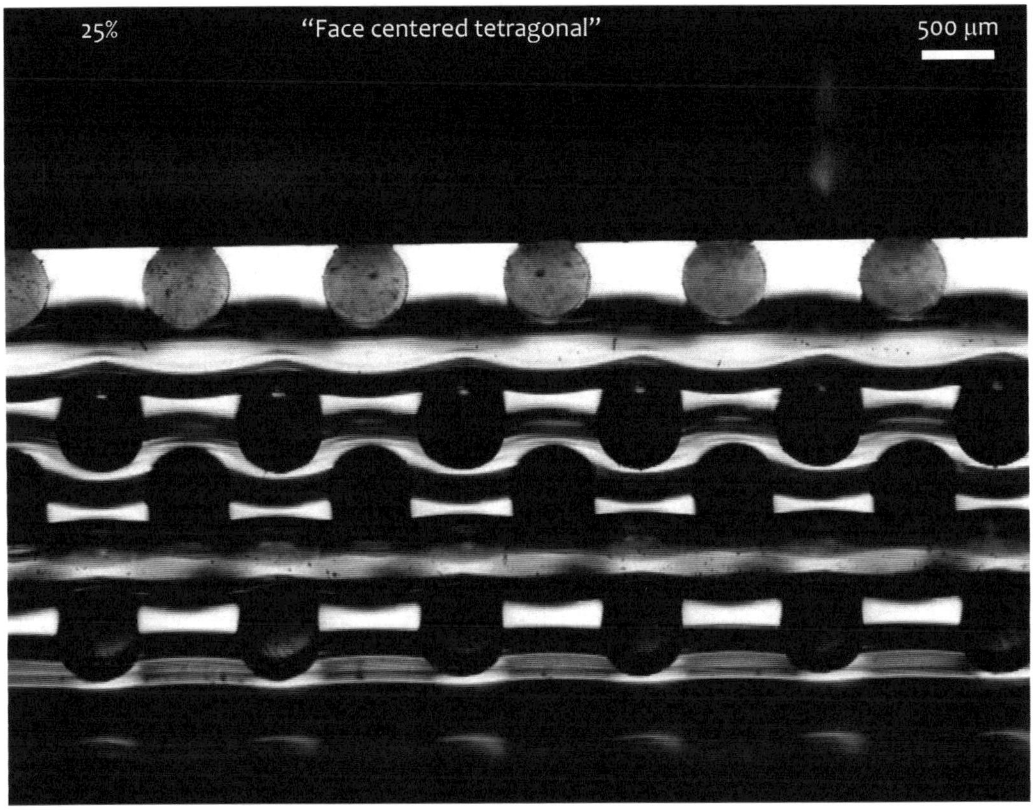
25%
"Face centered tetragonal"
500 µm

25% “Face centered tetragonal”

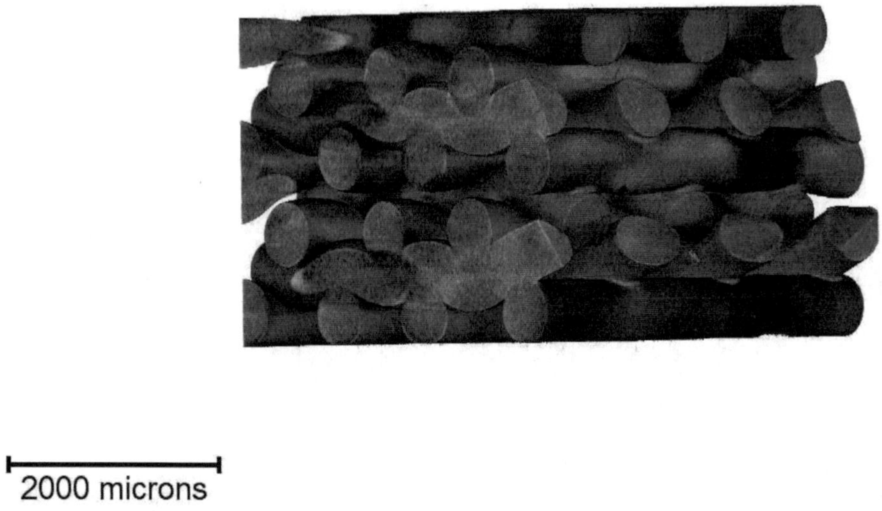

Finite element model of stress iso-surfaces: SC structure

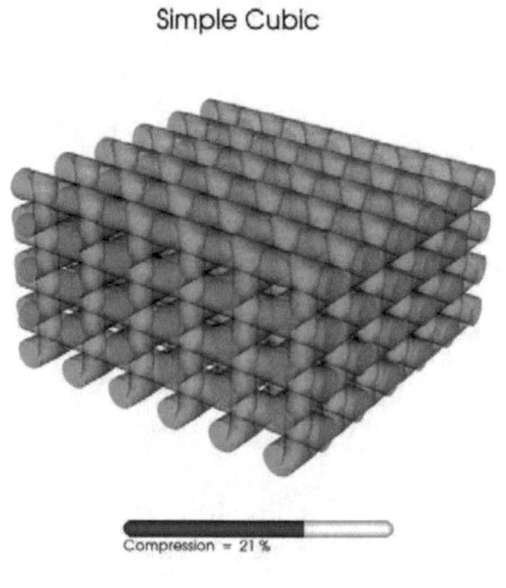

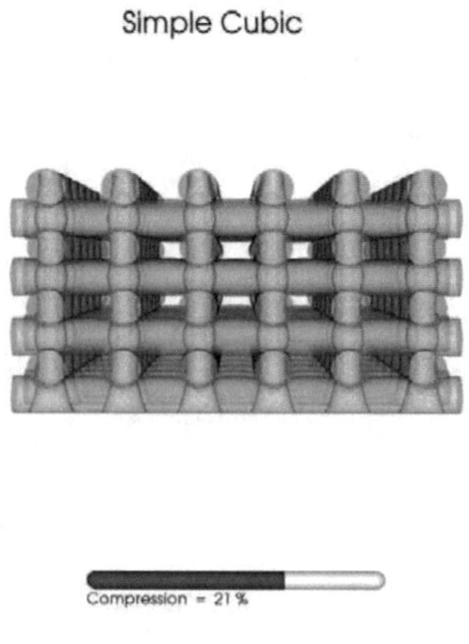

Lawrence Livermore National Laboratory

32

Finite element model of stress iso-surfaces: FCT structure

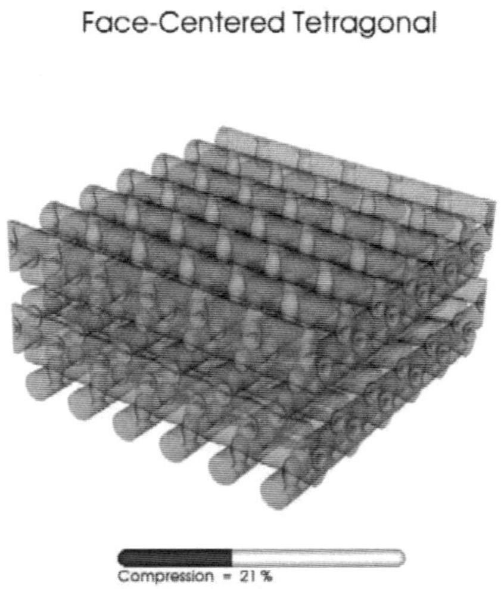

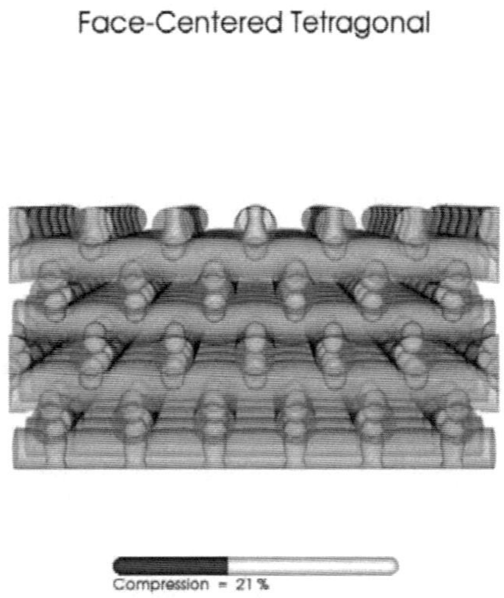

Lawrence Livermore National Laboratory

33

Architecture-dependent shear behavior under pre-compression: SC structure shows negative shear stiffness

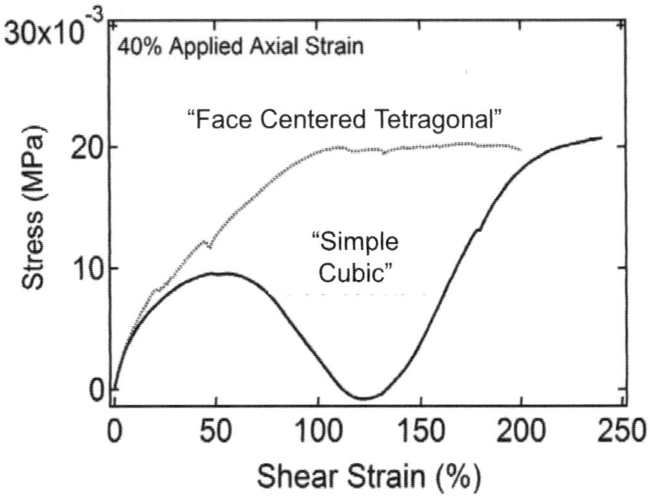

"Centered Tetragonal"

"Simple Cubic"

Lawrence Livermore National Laboratory

34

Towards commercialization of 3D printing

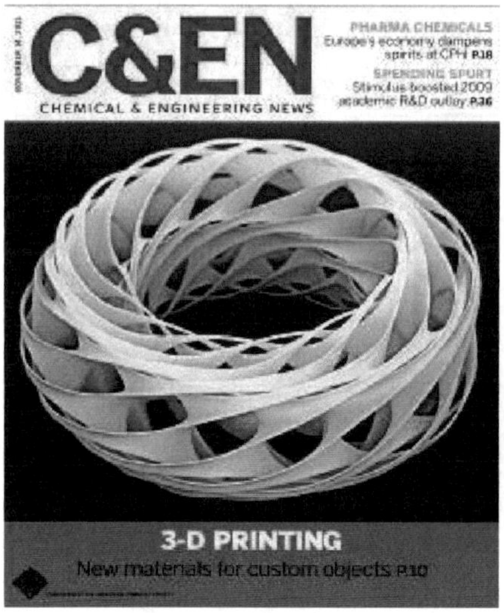

"Before personal manufacturing revolution can take place, though, researchers will need to develop a broader array robust printing materials ..."

Chemical & Engineering News, Nov 14, 2011 issue

Most 3D printing methods lack one or more the following attributes:

(1) *Materials flexibility;*
(2) *Ability to pattern fine features;*
(3) *Ability to fabricate more complex architectures;*
(4) *High throughput;*
(5) *High-speed printing.*

Lawrence Livermore National Laboratory

35

Thank you!

Lawrence Livermore National Laboratory

Control of microstructural architecture offers the potential for designed material properties and new functionalities

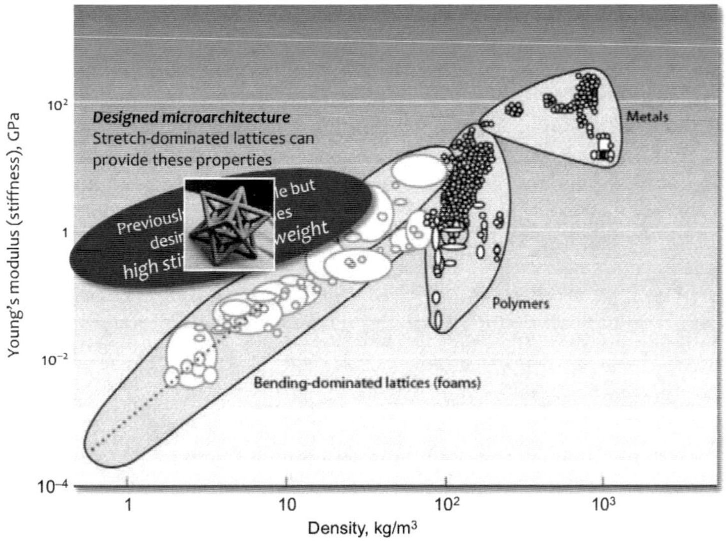

Order-of-magnitude performance improvement and decoupling of material properties is possible

Ink rheological properties: Non-Newtonian flow behavior and long print times

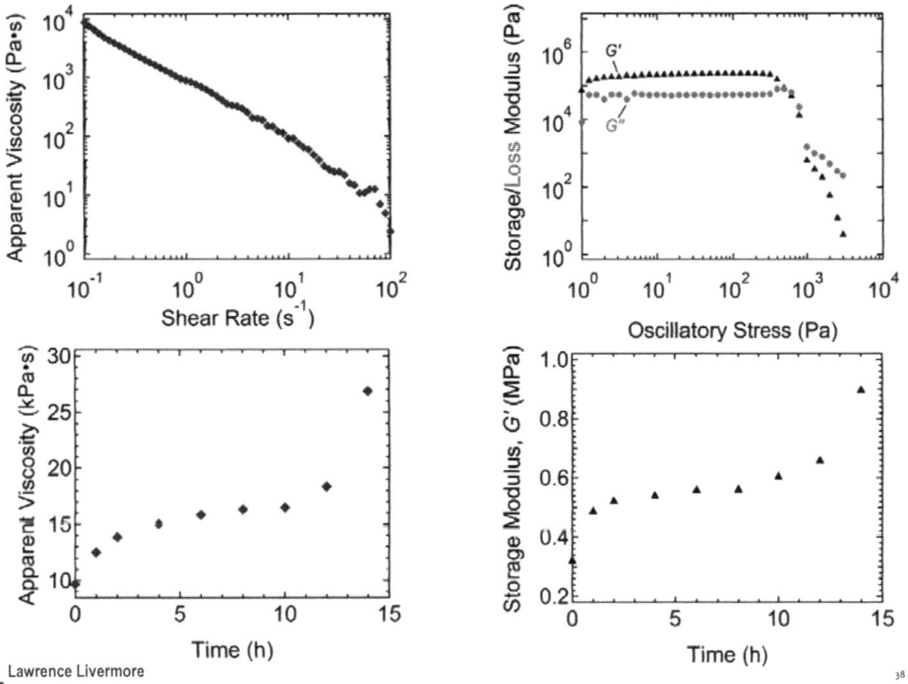

Printed structures will maintain their shape during thermal cure

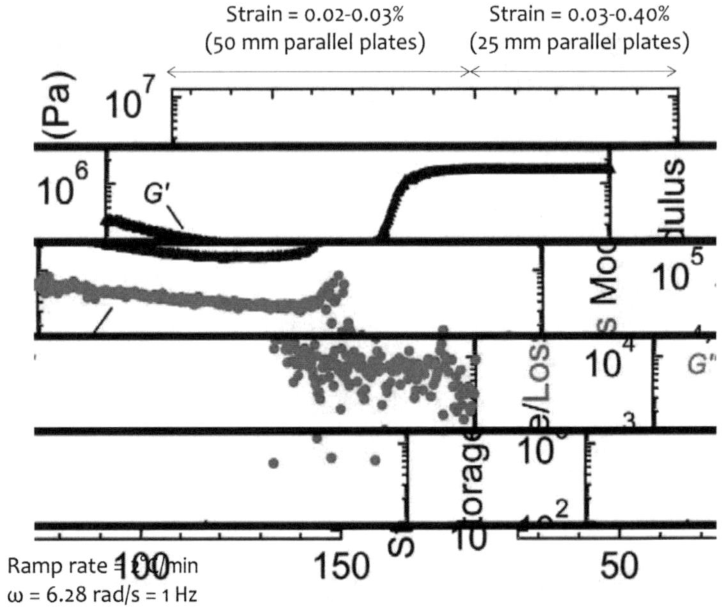

Lawrence Livermore National Laboratory

Future works: 3D microstructures have been theoretically derived for potential applications… but have not been built

Thermal Expansion

Steeves, C.A., et al., *J. Mech. and Phys. of Solids*, (2007)

Poisson's Ratio

Lakes, *Science*, (1987)

Microfluidics

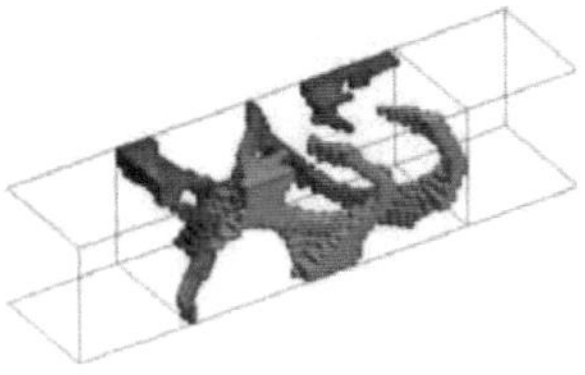

Sigmund, et. al., *Int. J. Numer. Meth. Fluids* (2009)

Piezoelectricity

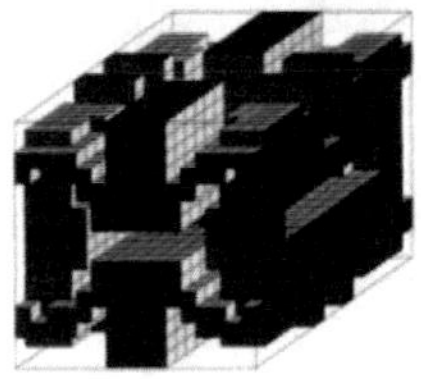

Sigmund & Torquato, *Smart Mater. Struct.* (1999)

Electrical & Thermal Conductors

Torquato, *et al.*, *PRL* (2002)

Photonic Crystals

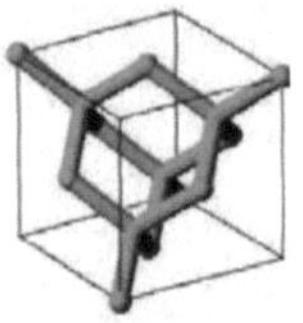

Maldovan & Thomas, *Nature Mater.* (2004)

Antennas

Erentok & Sigmund, *IEEE Antennas Propag.*, (2008)

Metamaterials

Zhou, *et al.*, IEEE Trans. Microwave Theory Tech. (2010)

Although these novel structures can be conceived and designed, they cannot currently be fabricated with the appropriate materials and size scales

Lawrence Livermore National Laboratory

Agenda

- Company Overview
- Definitions
- Project Development
- Process Development
  - Deposit Characterization - Sampling
  - Mine to Mill
  - Process Development
- Typical Test Work
- Summary

FLUOR.

2

Company Overview

FLUOR

3

Company Overview

- One of the world's leading publicly traded engineering, procurement, construction, maintenance, and project management companies
- **#110** in the FORTUNE 500 in **2013**
- Over **1,000** projects annually, serving more than **600** clients in **79** different countries
- **38,000** employees executing projects globally
- Offices in **30** countries on **6** continents
- Celebrated **100** years in **2012**

- 57 Years on NYSE
- Revenue: **$27.4 billion** (2013)
- New awards: **$25.1 billion** (2013)
- Backlog: **$34.9 billion** (2013)

FLUOR

4

Corporate Organization

ENERGY & CHEMICALS

- Biofuels
- Carbon Capture
- Chemicals & Petrochemicals
- Gas Processing & Gas Treating
- Gasification, Gas to Liquids/Chemicals, & IGCC
- Heavy Oil Upgrading & Oil Sands
- Hydrocarbon Transportation - Pipelines
- Liquefied Natural Gas (LNG)
- Offshore Oil & Gas Production
- Onshore Oil & Gas Production
- Petroleum Refining
- Polysilicon
- Sulfur Recovery
- Utilities & Offsites

POWER

- Environmental Compliance
- Gas-fueled/IGCC
- Nuclear
- Power Services
- Renewable Energy
- SMR Technology
- Solid-fueled
- Transmission

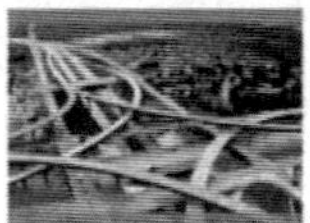

INFRASTRUCTURE

- Aviation
- Bridges
- Commercial & Institutional
- Offshore Wind Farms
- Ports & Marine Terminals
- Public-Private Partnerships
- Rail & Transit
- Telecommunications
- Toll Roads & Highways

MINING & METALS

- Mining
- Mining Process Expertise
- Metals
- Metals Process Expertise

INDUSTRIAL

- Life Sciences
- Manufacturing
- Operational Readiness
- Operations & Maintenance
- Water

GOVERNMENT

- Contingency Operations
- Design-Build
- Disaster & Emergency Response
- Nuclear Operations
- Nuclear Remediation
- Services & Base Operations
- Technical Support Services

FLUOR.

5

Worldwide Office Locations

Americas

Aliso Viejo, California
Anchorage, Alaska
Arlington, Virginia
Austin, Texas
Baton Rouge, Louisiana
Buenos Aires, Argentina
Calgary, Alberta, Canada
Caracas, Venezuela
Charlotte, North Carolina
Clarksville, Tennessee
Dallas, Texas
Greenville, South Carolina
Houston, Texas
Lima, Peru
Long Beach, California
Mexico City, Mexico
Port of Spain, Trinidad
Richland, Washington
San Francisco, California
San Juan, Puerto Rico
São Paulo, Brazil
Santiago, Chile
Vancouver, B.C., Canada
Washington, D.C.

Europe/Africa/Middle East

Abu Dhabi, U.A.E.
Ahmadi, Kuwait
Al Khobar, Saudi Arabia
Antwerp, Belgium
Asturias, Spain
Bergen-op-Zoom, The Netherlands
Dublin, Ireland
Farnborough, England
Gliwice, Poland
Haarlem, The Netherlands
Johannesburg, South Africa
London, England
Madrid, Spain
Moscow, Russia
Rotterdam, The Netherlands
Tarragona, Spain

Asia/Australia

Beijing, China
Brisbane, Australia
Cebu, Philippines
Jakarta, Indonesia
Manila, Philippines
Melbourne, Australia
New Delhi, India
Perth, Australia
Seoul, South Korea
Shanghai, China
Singapore
Tokyo, Japan

Years of Experience in Region

North America	South America	Europe	Africa	Middle East	Asia	Australia
103	76	63	53	66	62	63

FLUOR.

6

Comprehensive EPCM & PM Services

Life Cycle for Client Capital Assets

Design → Engineering → Procurement → Construction → Start-up → Operations & Maintenance → Client Solutions

Design

- Conceptual Design
- Estimating
- Feasibility Studies
- Permitting
- Process Simulation
- Project Financing
- Routing
- Scope Definition
- Siting
- Technology/ License Evaluation

Engineering

- Cost Control
- Detailed Engineering
- Fabrication
- Front-end Engineering
- Modular Construction
- Planning & Scheduling
- Process Simulation
- Safety Planning
- Systems Integration

Procurement

- Contract Staffing
- Contracts Management
- Equipment, Tools, and Fleet Services
- Expediting
- Fabrication
- Logistics
- Low Cost Country Sourcing
- Materials Management
- Purchasing
- Requirements Planning
- Sourcing
- Supplier Quality
- Warehousing

Construction

- Construction Management
- Contractor Management
- Craft Staffing & Training
- Equipment & Tools Supply
- Fabrication
- Field Mobilization
- Material Control
- Modular Construction
- Project & Program Management
- Quality Control
- Safety Programs

Start-up

- Commissioning
- Engineering Support
- Initial Production
- Plant Readiness
- Precommissioning
- Systems Checkout
- Turnover
- Validation

Operations & Maintenance

- Asset Performance Improvement
- Contract Staffing
- Equipment, Tools, and Fleet Services
- Facility Management
- Operational Readiness
- Plant Engineering
- Plant Operations & Maintenance
- Small Capital Projects
- Turnarounds, Outages & Shutdowns

FLUOR.

7

Fluor's San Francisco Office

- Fluor's San Francisco office was originally established by M&LS in 1996. Mining and Metals joined in September 2006 and relocated to our current location in 2008
- We offer full Engineering, Procurement and Construction Management services
- M&M group is experienced in a diverse range of mining projects and mineral processing technologies including:
  - Design of tailing systems and slurry pipelines
  - Technical experts recognized as industry leaders in critical areas, such as materials handling, comminution, mineral processing, hydrometallurgy, pyrometallurgy and waste treatment
- M&M focuses on the following mining markets in the Nevada/US Region:
  - Copper
  - Gold
  - Silver
  - Potash
  - Molybdenum
  - Iron Ore

Fluor San Francisco office

Current staff – approx. 100

- Industrial Services group expertise:
  - Design of facilities for Biotechnology, Microelectronics and Biofuels
  - Technologies include: Fermentation, Purification, Aseptic Fill & Finish
  - Specialized in cost effective execution of small retrofit projects
- IS focuses on the following markets:
  - Biotechnology
  - Microelectronics/ Photovoltaic
  - Biofuels

FLUOR.

8

Fluor's San Francisco Office
Project Highlights

- **Mining & Metals**
  - Allied Nevada, Hycroft Expansion Project EP
  - KUC, Cornerstone Expansion Project EPCM
  - KUC, Eagle Nickel Expansion Project, EPCM
  - FMI, Cu-Mo Thickener Replacement and Filter Installation Project, EP & CS
  - FMI, Sierrita Flotation Replacement Project, CS
  - Grupo Mexico, Cananea Project FS
  - Votorantim Metais, Votorantim Project, PFS
  - Barrick, Sustaining Capital Alliance, EPCM

- **Industrial Services**
  - Amgen, Bothell, Washington, Bothell Development Laboratory, Detailed Design
  - Amgen, Thousand Oaks, CA, B23 Facility Design Basis Studies
  - Bayer Healthcare, Berkeley, CA, Multi-product Campaign Feasibility Studies
  - Bayer CropSciences, West Sacramento, CA, Biopesticide Pilot Plant, Concept & Basic Engineering
  - Baxter, Hayward, CA, Multiple Manufacturing Support Studies
  - Genentech, Vacaville, CA, CCP1 Seed Lab Conceptual Design and Airlock Study
  - Genentech, South San Francisco, CA, CCP1 GMP Warehouse Detail Design
  - Polaris, DRX, Vacaville, CA, Conceptual Design
  - Polaris, TDW HBSP, Chundu, China, Conceptual Design

FLUOR.

9

Definitions

FLUOR.

10

Definitions

- Metallurgist/Metallurgical Engineer – a process engineer specializing in the separation and extraction of minerals and or metals form a raw material. Typically have a degree in Chemical, Metallurgical, Mineral Processing Engineering
- Metallurgical Performance – refers to grade (quality) and recovery (yield)
- Feasibility Studies
  - The most formal and reliable study is the feasibility study. This study requires application of Society of Mining Engineers ("SME") standards, actual vendor quotations for the capital and operating cost parameters, and detailed analysis of the testing and engineering. Such a study will often provide a reliability of +/- 10 to 20%.15 Simply put, the feasibility study is a formal technical report that is used by the company to determine whether the proposed project is capable of being developed at a sufficient return to justify the capital and managerial resources that must be committed.

FLUOR.

11

Definitions

- A 'Mineral Resource' is a concentration or occurrence of solid material of economic interest in or on the Earth's crust in such form, grade (or quality), and quantity that there are reasonable prospects for eventual economic extraction. The location, quantity, grade (or quality), continuity and other geological characteristics of a Mineral Resource are known, estimated or interpreted from specific geological evidence and knowledge, including sampling. Mineral Resources are sub-divided, in order of increasing geological confidence, into Inferred, Indicated and Measured categories.
- An 'Ore Reserve' is the economically mineable part of a Measured and/or Indicated Mineral Resource. It includes diluting materials and allowances for losses, which may occur when the material is mined or extracted and is defined by studies at Pre-Feasibility or Feasibility level as appropriate that include application of Modifying Factors. Such studies demonstrate that, at the time of reporting, extraction could reasonably be justified. The reference point at which Reserves are defined, usually the point where the ore is delivered to the processing plant, must be stated. It is important that, in all situations where the reference point is different, such as for a saleable product, a clarifying statement is included to ensure that the reader is fully informed as to what is being reported.

FLUOR.

12

Project Development

FLUOR

13

Project Development – Process Involvement

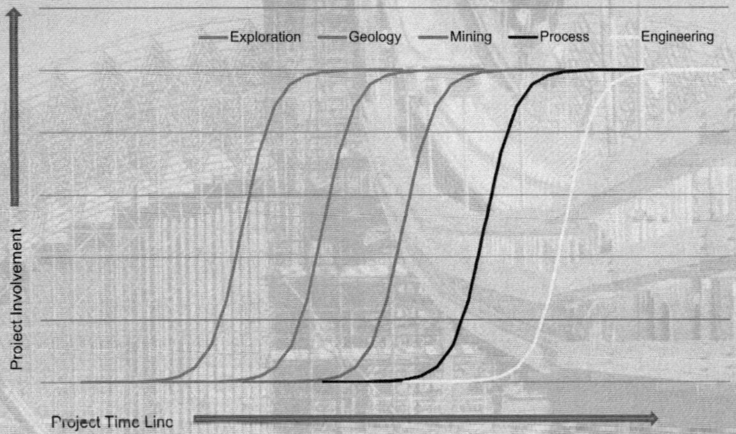

FLUOR

14

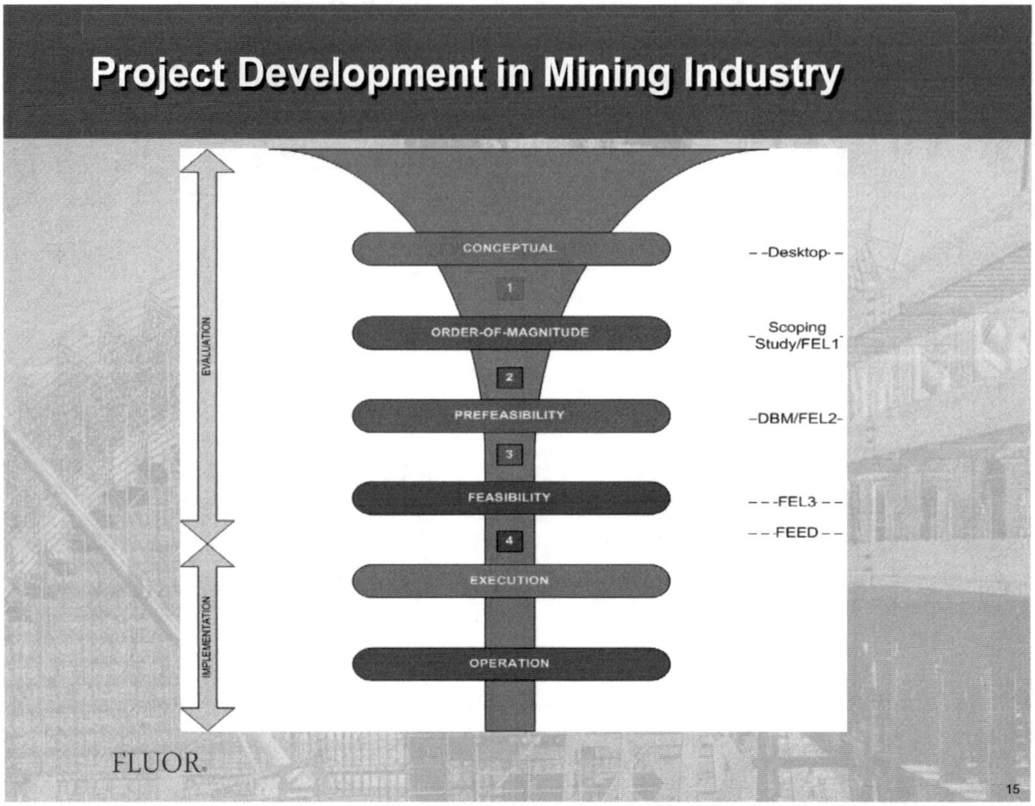
Project Development in Mining Industry
EVALUATION
IMPLEMENTATION
CONCEPTUAL
1
ORDER-OF-MAGNITUDE
2
PREFEASIBILITY
3
FEASIBILITY
4
EXECUTION
OPERATION
--Desktop--
Scoping Study/FEL1
-DBM/FEL2-
---FEL3--
---FEED--
FLUOR
15

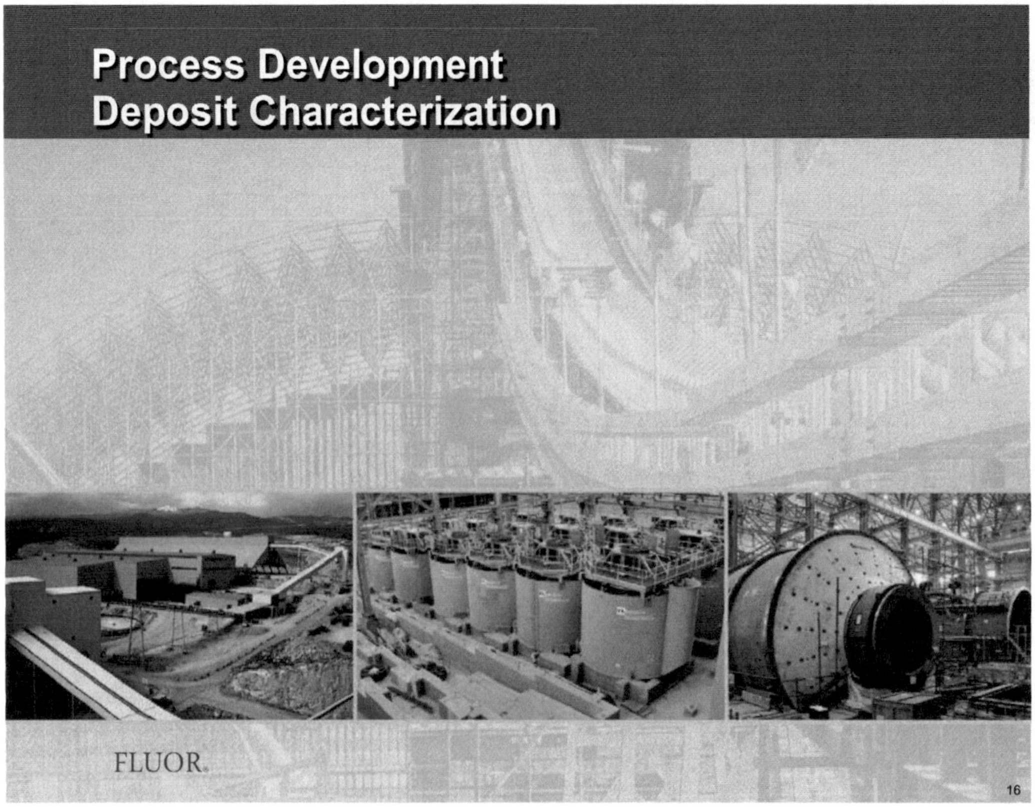
Process Development
Deposit Characterization
FLUOR
16

Exploration – The Starting Point

- Mineral exploration is the process of meticulously searching for and analyzing potential areas where useful or precious minerals, metals, and other resources can be mined or otherwise extracted from the earth. Mining companies may conduct mineral exploration projects, akin to highly advanced prospecting, to learn more about potential sites for new mines

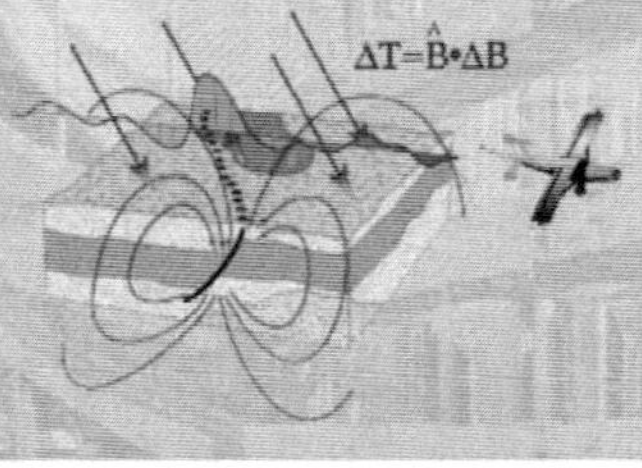

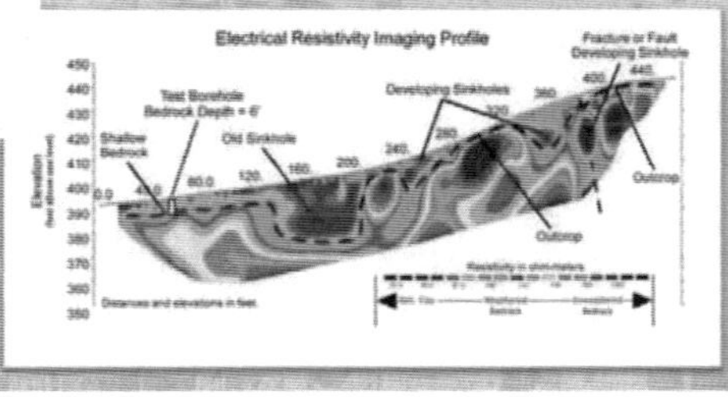

FLUOR.

17

Exploration

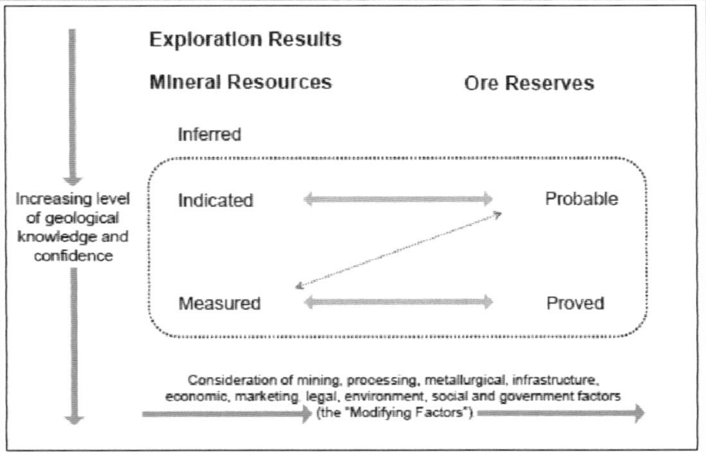

Source: Australasian Code for Reporting of Exploration Results, Mineral Resources and Ore Reserves, The JORC Code 2012

FLUOR.

18

Sampling

- To determine the physical and chemical characteristics of materials associated with a potential ore body
- Types of Sampling
  - Core Drilling
  - Reverse Circulation Drilling
  - Auger Drilling
  - Channel Sampling
  - Trench Sampling
  - Grab Sampling

- Types of Analysis
  - In-Situ Measurements
  - Geological Logging
  - Chemical Analysis
  - Mineralogical Assessment
  - Physical Characteristics
  - Metallurgical Testing

FLUOR.

19

Drill Hole Location Mapping

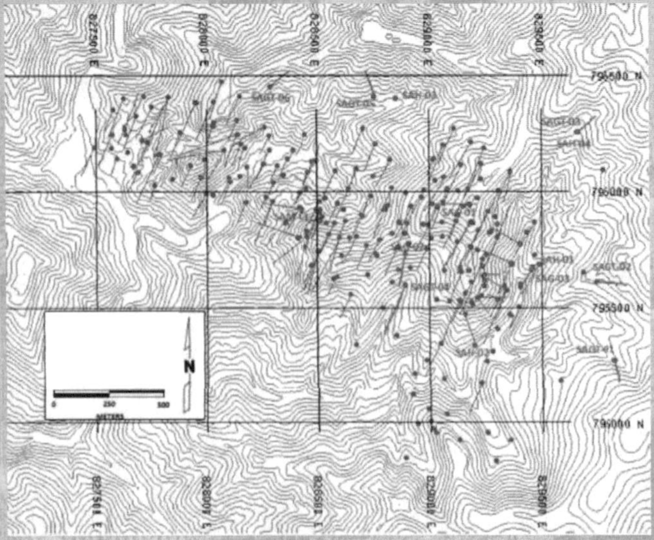

Source: www.sedar.com

FLUOR.

20

Chemical Analysis - Grade

- Representative portions of samples are taken for chemical analysis.
- Analysis can include
  - Moisture Content
  - Total Metal Content
  - Soluble Metal Content
  - Non-metal components
  - Heat content (coal)

FLUOR.

21

Chemical Analysis – Grade & Location

- Core Model
  - http://www.corebox.net/properties/escobal

FLUOR.

22

Chemical Analysis - Grade

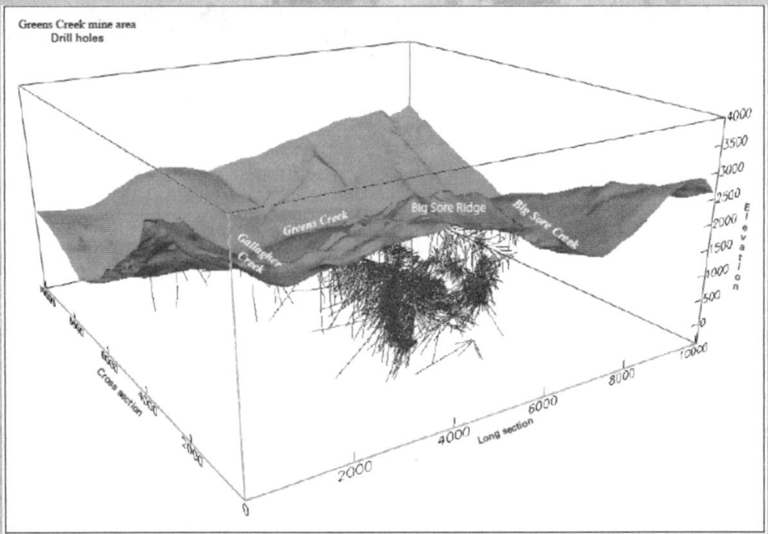

Source: USGS, **Three-Dimensional Modeling and Visualization of Greens Creek Drill-Hole Data,** By Gregory K. Lee and Cliff D. Taylor Chapter 14 of **Geology, Geochemistry, and Genesis of the Greens Creek Massive Sulfide Deposit, Admiralty Island, Southeastern Alaska** ,Edited by Cliff D. Taylor and Craig A. Johnson, Professional Paper 1763

FLUOR.

23

Chemical Analysis - Grade

Isovalue shell display of zinc greater than 10 percent in orebodies, Greens Creek mine.

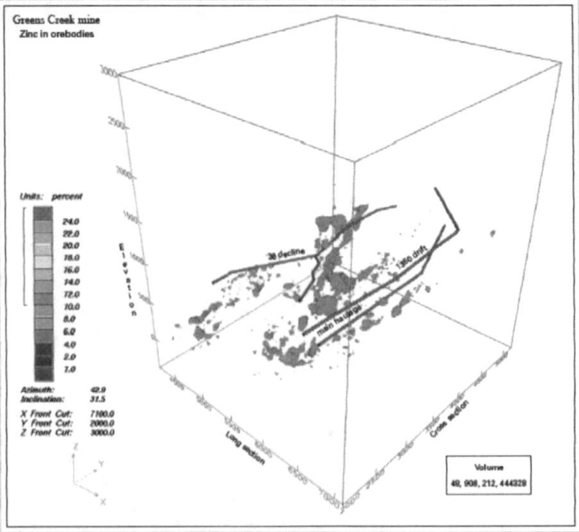

Source: USGS, **Three-Dimensional Modeling and Visualization of Greens Creek Drill-Hole Data,** By Gregory K. Lee and Cliff D. Taylor Chapter 14 of **Geology, Geochemistry, and Genesis of the Greens Creek Massive Sulfide Deposit, Admiralty Island, Southeastern Alaska** ,Edited by Cliff D. Taylor and Craig A. Johnson, Professional Paper 1763

FLUOR.

24

Chemical Analysis - Grade

Isovalue shell display of arsenic: antimony ratios greater than 10 in orebodies, Greens Creek mine.

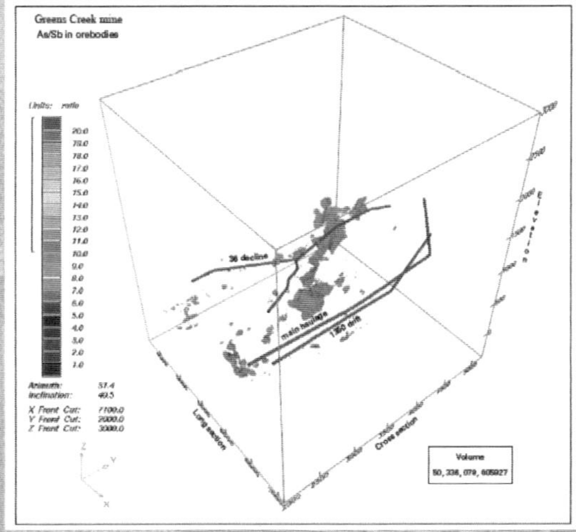

Source: USGS, **Three-Dimensional Modeling and Visualization of Greens Creek Drill-Hole Data,** By Gregory K. Lee and Cliff D. Taylor Chapter 14 of **Geology, Geochemistry, and Genesis of the Greens Creek Massive Sulfide Deposit, Admiralty Island, Southeastern Alaska** ,Edited by Cliff D. Taylor and Craig A. Johnson, Professional Paper 1763

FLUOR.

25

Mineralogical Assessment

- Many metallurgical processes act to separate minerals contained in ores
- Amounts of minerals can be determined visually, or with direct measurements using SEM
- The SEM images can be used to provide information on
  - Modal abundance
  - Grain size
  - Liberation characteristics
  - Initial projections of recovery

FLUOR.

26

Mineralogical Assessment

Chalcopyrite
$CuFeS_2$

Source: Wikipedia

Cuprite
Cu_2O

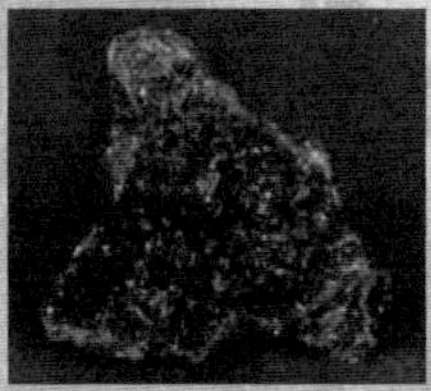

Source: Wikipedia

Covellite
CuS

Source: Wikipedia

FLUOR.

AV\20020429037.ppt 27

Mineralogical Assessment

FLUOR.

28

Physical Characteristics

- Direct Measurement of the physical parameters of the samples is also performed, such as:
  - In-situ density
  - Porosity
  - Permeability
  - Compressive strength
  - Compaction
  - Grindability

FLUOR

29

Geometallurgy

- Historically most of the sampling and selection of samples has generally been performed by Geologists with a focus on resource delineation
- Little thought went into selection of samples for metallurgical test work
- Can lead to test results that are of little to no use to the design engineer.
- Geometallurgy is a scientific discipline in which geological data, mining data and processing data are coanalysed to generate useful information and knowledge to optimize resource profitability.

FLUOR

30

Geometallurgy & Sample Selection

- Implications of Geometallurgy – Metallurgical Test Work has to address variability in ore types, for example:
  - Composite 1 - Copper Grade in the 0.2 to 0.3 Percent Range
  - Composite 2 - Chalcopyrite Dominant ≥ 0.2 to 0.5 Percent Copper
  - Composite 3 - Chalcopyrite Dominant ≥ 0.5 Percent Copper
  - Composite 4 - Bornite Moderate to Strong ≥ 0.2 to 0.5 Percent Copper
  - Composite 5 - Bornite Moderate to Strong ≥ 0.5 Percent Copper
  - Composite 6 - High Copper Grade
  - Composite 7 - Mid Copper Grade
  - Composite 8 - Low Copper Grade
  - Composite 9 - SE Low Grade
  - Composite 10 - SE Mod-High Grade
  - Composite 11 - SE High Bornite
  - Composite 12 - SW Low Copper Grade
  - Composite 13 - SW Moderate High

FLUOR.

31

Process Development Mining

FLUOR.

32

Mining

- Mine Planning - is complex and integrates several technical branches of the mining discipline including rock mechanics, rock breakage, ventilation and mine design.
- The overriding factor in the success of a mining operation and what ultimately determines whether it will proceed through feasibility studies and onto construction and development is the level of profitability. The drive to exploit a mineral resource such that it adds as much value as possible to the mining business requires a continual update of the mine plan and the rapid implementation of any value adding opportunities as they arise.

FLUOR

33

Mining

- Block Models and Floating Cones

Block	1	2	3	4	5	6	7	8	9
1	-3	-3	-3	-3	-3	-3	-3	-3	-3
2	-4	7	-4	6	-4	-4	-4	-8	-4
3	-5	-5	-5	15	-5	-5	-5	-5	-5

Stage	Block No	Block Value	Cone Value	Minable
1	(3,4)	15	-2	No

- Process Involvement
  - Economic value of a block reflects contained value
  - Also accounts for estimated recoveries and operating costs
  - Must reflect proposed processing route
  - Iterative process to optimize final mine plan
  - Test work focus on early years of mine production

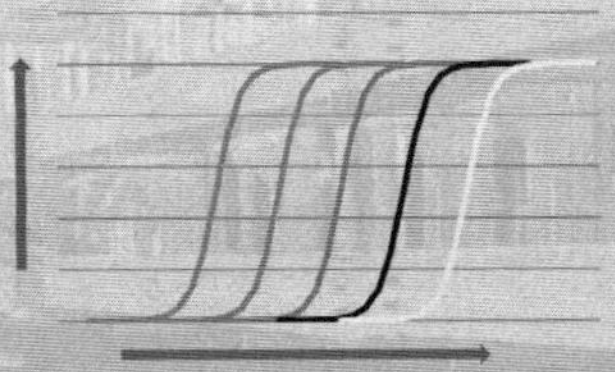

FLUOR

34

Process Development
Process Engineering

FLUOR.

35

Process Development in Mining Industry

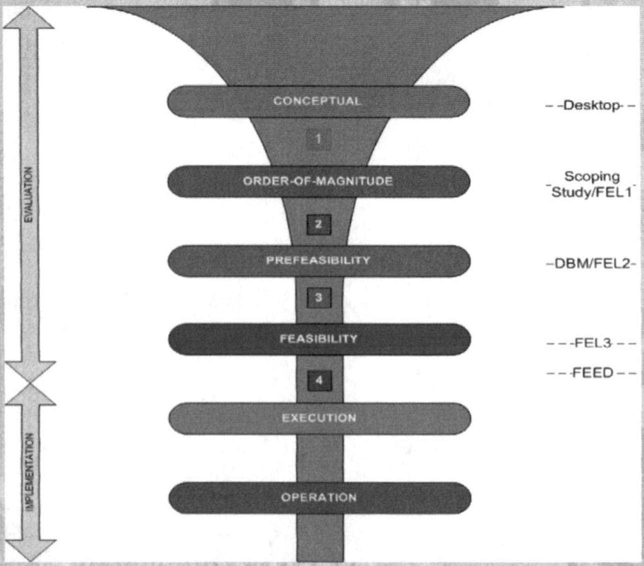

FLUOR.

36

Process Development – Conceptual

- Available information
  - Potential flow sheets identified based on 'typical' industry practice
  - Processing characteristics assumed from comparable operations
  - General awareness of advances in technology that may be applicable
  - High level understanding of water balance

Low Impact Crusher Test
Courtesy Rob Sloan, ALS Minerals

- Objectives
  - Initial assessment of plant (minimal engineering)
  - Equipment selection is factored from similar operations
  - Initial production schedules tied to preliminary mining plans
  - Metallurgical performance based on analogous operations

FLUOR.

37

Process Development – Order-of-Magnitude

- Available Information
  - Preliminary assessment of processing options and throughput
  - Preliminary process data being produced
  - Sampling requirements identified in line with processing option
  - Laboratory results and modeling supplemented by literature search to support proposed processing option
  - Material handling, processing and tailings stream considered
  - Potential product contaminants identified
  - Information compiled on new, potentially applicable technologies
  - Preliminary assessment of water quality and quantity made and basic water balance modeling completed

FLUOR.

38

Process Development – Order-of-Magnitude

- Objectives
  - Reviewed physical, metallurgical and processing plant requirements for each option
  - Major equipment identified consistent with test work information
  - Long lead items identified
  - Preliminary throughput identified
  - Preliminary plant design criteria defined
  - Summary level PFD an H&MB prepared

Drop Weight Test
Courtesy Rob Sloan, ALS Minerals

FLUOR.

39

Process Development – Prefeasibility

- Available Information
  - Acquire a representative suite of samples to support test work program (may reflect proof of concept test, pilot testing, environmental, etc.)
  - Critical parameters, grade-recovery, extraction-recovery, have been determined with reasonable assurance
  - Consider options and select preferred flow sheet
  - Physical properties reflected in design
  - Optimum reagents identified
  - All deleterious trace elements accounted for
  - New technologies assessed for reliability, robustness, extent of application, and cost

Batch Autoclave Test
Courtesy Nik Bogdanovic, Fluor

FLUOR.

40

Process Development – Prefeasibility

- Objectives
  - Detailed justification to support chosen flow sheet and elimination of other options
  - Major process steps and equipment, product transport, utility requirements described
  - Complete detailed specifications for major equipment and duty specifications for minor equipment
  - "Non-standard" equipment identified
  - Detailed plant design criteria defined
    - Product specifications
    - Consumption rates for operating and maintenance consumables
    - Blending and stockpiling strategy
    - Detailed plant layouts

FLUOR

41

Process Development - Feasibility

- Available information
  - Design Criteria (all disciplines), equipment specifications, raw materials identified
  - Supplementary process data to address off-design consideration and/or process uncertainties
  - Test work has confirmed final plant productivity, final recovery rates, operating parameters, and process optimization issues
  - All reagents identified and optimized

Pilot Autoclave
Courtesy Rod Clary, Fluor

FLUOR

42

Process Development - Feasibility

- Objectives
  - Engineering designs are developed to a level suitable for supporting required accuracy of cost estimate (AACE Class 2)
  - Detailed equipment list, load lists, identification of preferred suppliers
  - Detailed production schedule (by monthly for first 2 years)
  - Process route is "frozen"
  - PFD's finalized
  - H&MB complete for processes and services
  - Basic design for tailings pond design complete.

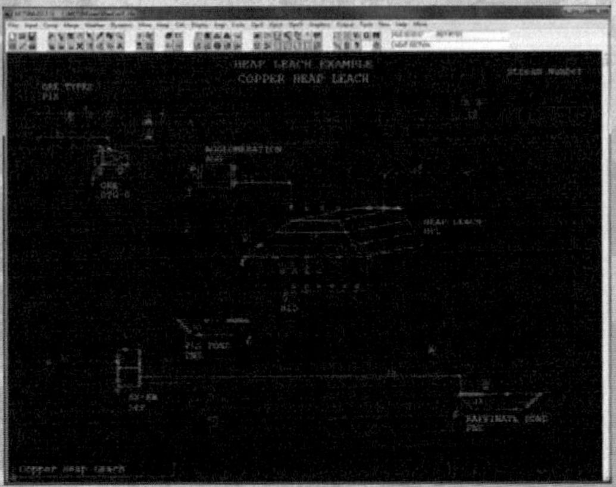

FLUOR.

43

Environmental Considerations

- Water and Water Management
  - Current trends to maximum recycle of water
  - Climatic conditions are key
- Tailings Disposal
  - Cyanide destruction technologies are well understood
  - Ore bodies are becoming more complex, processing techniques are advancing to manage 'bad actors" e.g. arsenic.
- Permitting
  - Process supports permitting throughout project life cycle

FLUOR.

44

Typical Test Work

FLUOR

45

Process Development – Typical test work

- Conceptual
  - Crushing and Grinding
  - Batch Flotation Tests
  - Cyanidation Bottle Roll Tests
  - Leaching/Extraction test work
  - Solid Liquid Separation Tests

Cone Crusher
Courtesy Nik Bogdanovic, Fluor

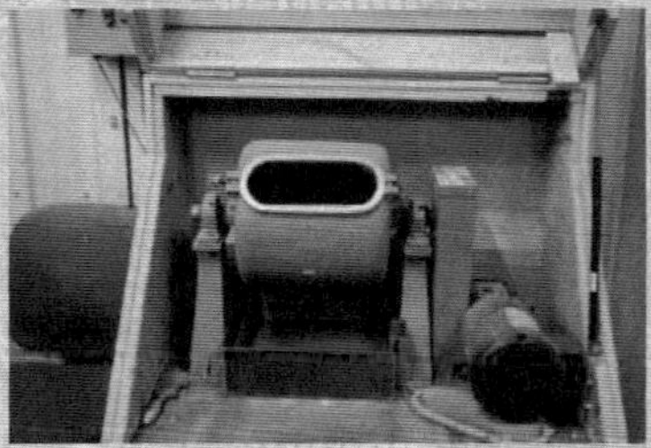

Bond Work Index Test
Courtesy Rob Sloan, ALS Minerals

FLUOR

46

Process Development – Typical Test Work

- Pre-Feasibility/Feasibility
  - Proof of Concept
  - Pilot Testing
  - Effect of recycle water
  - Rheological testing
  - Environmental
  - Reagent Optimization

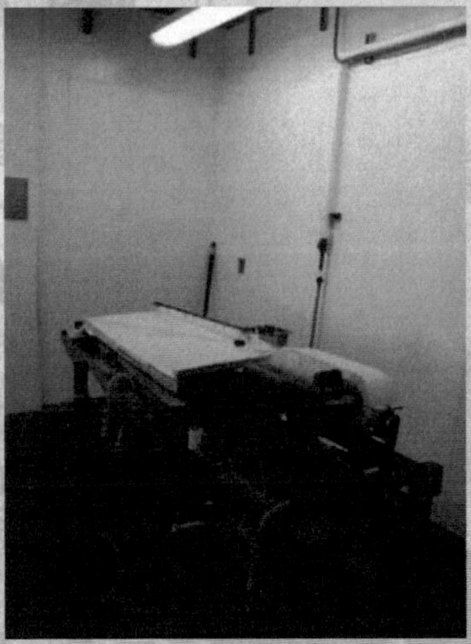

Gravity Concentration
Courtesy Nik Bogdanovic, Fluor

FLUOR

47

Process Development – Typical Test Work

- Test Work
  - Lab / batch
  - Lab /Continuous
  - Micro Plant
  - Pilot plant
  - Demonstration Plant

Source: Internet – Demonstration Plant

Source: www.sgs.com, "Bankable Pilot Plant – Fully Integrated Simulations of Continuous Operating Conditions"

FLUOR

48

Summary

- Process Development in Mining Industry is a collaborative effort of multiple disciplines.
- Key is to understand what is being delivered to the processing facilities and when (Ore Characterization & Mine Plan).
- Metallurgical Process Engineer needs to understand various aspects of the others work to make sound process decisions.

FLUOR.

49

CH2MHILL
Plateau Remediation Company

US DOE Technology Development Approach –

Application to Process Development for K-Basin Sludge Processing at the Hanford Nuclear Site

Presented by: Jim Honeyman
CHPRC Senior Technical Advisor

2014 NorCal Process Development Symposium

U.S. DEPARTMENT OF ENERGY

One Team. One Culture.

Background

- **DOE and its predecessor agencies**
  - Historically produced nuclear weapons and nuclear materials beginning in 1943
  - Processes and Facilities to treat and dispose of high hazard nuclear waste and environmental cleanup
- **Many production facilities first-of-a-kind; one-of-a-kind**
  - Complex chemical processes
  - Radioactive materials
  - Often remotely operated and maintained
- **Capital Decision Processes are intentionally deliberate and formalized**
  - DOE O 413.3B -- project management framework
  - Significant facility costs ($100 Millions - $Billions)
  - Significant public environmental risks and hazards
- **History of significant cost and schedule issues with radioactive materials facilities**
  - Short cuts in the design process ("fast-track" design prior to completion of supporting testing)
  - Commitment to cost and schedule estimates before a complete technical design basis established
  - Technology "favorites" or "not invented here" can drive technology selections
  - Overconfidence in the technical maturity of <u>similar</u> applications elsewhere

U.S. DEPARTMENT OF ENERGY

CH2MHILL
Plateau Remediation Company 2

Rand Corporation "Pioneer Plants" studies on cost/schedule growth; production performance

- Series of studies conducted between 1979-1986
  - Sponsored by DOE
- Evaluated 44 energy, chemical, petrochemical projects
  - categorized the primary drivers of production plant cost and schedule growth
  - Impacts to expected production performance

"Despite widespread belief to the contrary, unanticipated inflation, unanticipated regulatory changes, scope changes, and other external factors such as bad weather and strikes, are <u>not</u> the principal causes of cost and schedule underestimation."

U.S. DEPARTMENT OF ENERGY

CH2MHILL Plateau Remediation Company 3

Primary Causes of Schedule, Cost, Productivity Issues- Pioneer Plants Study

- **<u>The extent</u> that a plant's <u>technology</u> departs from that of prior plants**
- **The <u>degree of definition</u> of the project's <u>site</u> and related characteristics**
- **The plant's <u>process complexity</u> (number of unit operations/ number that represent a new application)**
- **And finally**
  - **Does the plant utilize <u>solid feedstock</u> and/or <u>solids transfers</u> in its process flowsheet?**

A key issue in process plant technology selection and process design, is how <u>technology maturity</u> is assessed within the context of the <u>proposed application</u> for a given project

U.S. DEPARTMENT OF ENERGY

CH2MHILL Plateau Remediation Company 4

Need for Technology Maturation Methodology

- **NASA and DOD -- Technology decisions historically relied on expert judgment**
- **NASA and DOD began the development of a more universal technology assessment methodology beginning in 1989.**
  - Developed Technology Readiness Assessment methodology (DoD Technology Readiness Assessment Deskbook, July 2009)
- **DOE adapted this methodology to DOE waste processing and treatment facilities/projects.**
  - DOE-EM conducted several pilot assessments and refined the methodology for application to radioactive processes and process technologies
  - DOE adopted the EM-TRA methodology in 2011 (DOE G 413.3-4A)
  - Integrated the TRA process into its formal Critical Decision Process (DOE O 413.3B)

U.S. DEPARTMENT OF ENERGY

CH2MHILL Plateau Remediation Company

5

Technology Readiness Assessment Guide/ Process

- **Defines Critical Technology Elements (CTE's) for a process**
- **Evaluates the size and scale of the proposed technology & extent that development/ demonstration for the application is completed**
  - **Utilizes a series of questionnaires to compare the technology element with expected status at each TRL**
- **Considers the programmatic considerations and the state of the supply chain**
  - **Safety, environmental permits, project planning, committed funding?**
  - **Can the required components be obtained at the size/scale needed and at the quality levels required?**
- **Provides a basis to compare technologies at differing levels of maturity for decision-making**
  - **helps to quantify maturity development remaining**

U.S. DEPARTMENT OF ENERGY

CH2MHILL Plateau Remediation Company

6

CTEs are evaluated and assigned a TRL value between 1-9

Relative Level of Technology Development	Technology Readiness Level	TRL Definition
System Operations	TRL 9	Actual system operated over the full range of expected mission conditions.
System Commissioning	TRL 8	Actual system completed and qualified through test and demonstration
	TRL 7	Full-scale, similar (prototypical) system demonstrated in relevant environment
Technology Demonstration	TRL 6	Engineering/pilot-scale, similar (prototypical) system validation in relevant environment
Technology Development	TRL 5	Laboratory scale, similar system validation in relevant environment
Technology Development	TRL 4	Component and/or system validation in laboratory environment
Research to Prove Feasibility	TRL 3	Analytical and experimental critical function and/or characteristic proof of concept
	TRL 2	Technology concept and/or application formulated
Basic Technology Research	TRL 1	Basic principles observed and reported

[1] Simulants should match relevant chemical and physical properties.

[2] Testing with as wide a range of actual waste as practicable and consistent with waste availability, safety, ALARA, cost and project risk is highly desirable.

U.S. DEPARTMENT OF ENERGY

CH2MHILL Plateau Remediation Company 7

DOE/EM TRL Testing Requirements

TRL Level	Scale of Testing	Fidelity	Environment[1,2]
9	Full	Identical	Operational (Full Range)
8	Full	Identical	Operational (Limited Range)
7	Full	Similar	Relevant
6	Engineering/Pilot Scale	Similar	Relevant
5	Lab/Bench	Similar	Relevant
4	Lab	Similar	Relevant
3	Lab	Pieces	Simulated
2		Paper	
1		Paper	

[1] Simulants should match relevant physical and chemical properties

[2] Testing with as wide a range of actual waste as practicable; and consistent with waste availability, safety, ALARA, cost, and project risk is highly desirable

U.S. DEPARTMENT OF ENERGY

CH2MHILL Plateau Remediation Company 8

Linkage of Project Decisions and Technology Readiness Levels

Mission Need CD-0 ⇨ Alternative Selection CD-1 ⇨ Performance Baseline CD-2 ⇨ Construction Start CD-3 ⇨ Operations Start CD-4

- CD-1: TRA 1 (TRL=4) TMP
- CD-2: TRA 2 (TRL=6)
- CD-3: TRA 3* (TRL=6)

Technology Requirements Review | Conceptual Design Review | Preliminary Design Review | Final Design Review | Operational Readiness Review

* TRA 3 required if there is technology modification/change on going from preliminary to final design.

U.S. DEPARTMENT OF ENERGY

CH2MHILL Plateau Remediation Company 9

BTF-TWRS Overview 4/7/99

U.S. DEPARTMENT OF ENERGY

CH2MHILL Plateau Remediation Company 11

Hanford 100K Area

U.S. DEPARTMENT OF ENERGY

CH2MHILL Plateau Remediation Company

K-Basin Layout

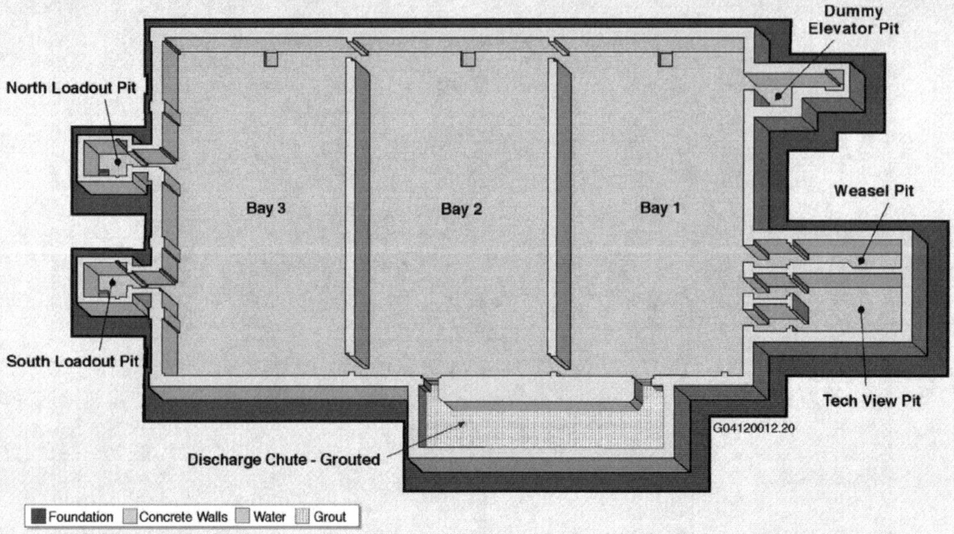

U.S. DEPARTMENT OF ENERGY

CH2MHILL Plateau Remediation Company

13

Working in the Basin at 100 KW

U.S. DEPARTMENT OF ENERGY

CH2MHILL Plateau Remediation Company

Sludge Description

- **Sludge consists of:**
  - Radionuclides - Corroded N Reactor fuel
    - Uranium Metal (dense, pyrophoric) and Uranium Oxides
  - Corrosion Products – Iron, aluminum, and uranium hydroxides
  - Environmental particulates – sand, rocks, organic media
  - Basin Operations – Spalled concrete, paint flakes , sand (sand filters), ion exchange materials
- **Particle size less than 0.25 inches; down to sub-micron sizes**
- **Specific Gravity ranges from ~1 to ~19; slurry transfer velocities >12 ft/sec**
- **Challenging non-newtonian slurries**
- **Approximately 30m³ total initial volume**
  - ~23m³ Container Sludge
  - ~5 m³ Settler Sludge
- **Dose Rates - 1 to 400 R/hr**
- **Designated as RH-TRU**

U.S. DEPARTMENT OF ENERGY 15 CH2MHILL Plateau Remediation Company

CH2M HILL Approach – beginning in 2008

- **Conducted Formal Alternatives Analysis -- proposed 2-phase strategy**
  - Phase 1 to remove the sludge, transport it away from the river to enable completion of D&D and waste site remediation along the river
  - Phase 2 will focus on future sludge treatment and packaging
- **In parallel -- conducted sludge sampling and characterization to establish technical and safety basis for design**
- **Planned a low technical and programmatic risk approach**
  - Apply TRA process and the DOE O 413.3B critical decision process
    - Developed a preliminary Technology Maturation Plan
      - Component, subsystem, and integrated testing were done in conjunction with the ongoing design
      - Determined that full scale, integrated demonstration would be required to demonstrate TRL-6 due to challenging solids characteristics and slurry transfers
      - Developed a non-radioactive simulant strategy that bounded the 3 different inventories of radioactive sludge.

U.S. DEPARTMENT OF ENERGY CH2MHILL Plateau Remediation Company 16

Engineered Container Retrieval and Transfer System Simplified Process Diagram

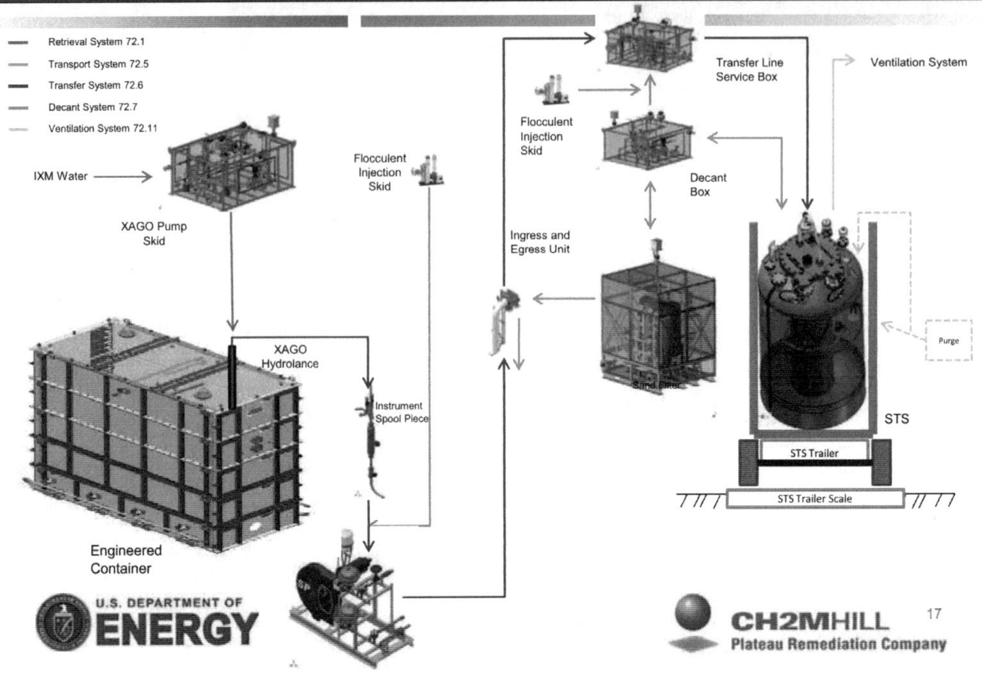

U.S. DEPARTMENT OF ENERGY

CH2MHILL Plateau Remediation Company

17

Integrated Waste Process System Testing Test Setup

A key component of the relevant environment that was replicated for the process equipment (i.e., XAGO™ HydroLance and Booster pump) was operating underwater with appropriate transfer hose lengths and elevation changes.

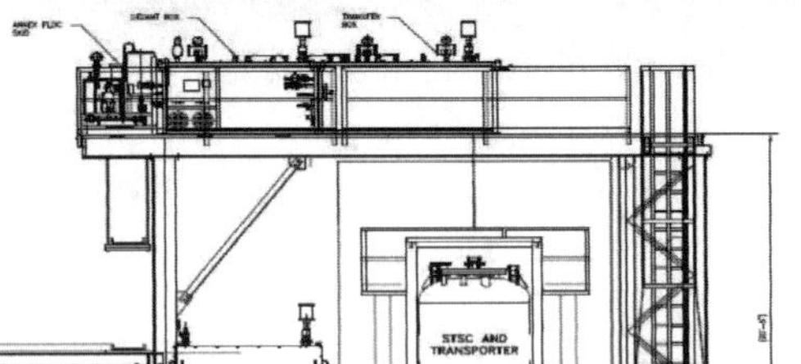

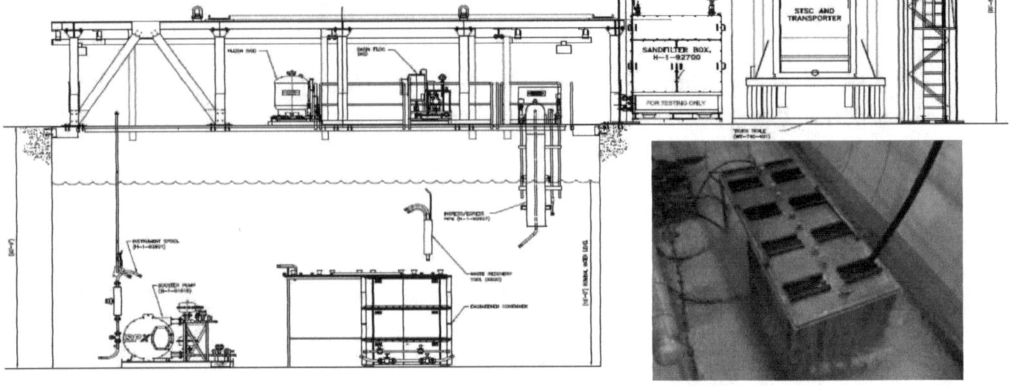

U.S. DEPARTMENT OF ENERGY

CH2MHILL Plateau Remediation Company

18

Integrated Waste Process System Testing

U.S. DEPARTMENT OF ENERGY

CH2MHILL Plateau Remediation Company 19

Technology Maturation Approach

- Conducted targeted component level testing to support conceptual design and technology selection
- Conducted a TRA Self-Assessment of the technology elements utilizing the TRL questionnaires – built documentation basis for TRL self determination
- DOE TRA team identified Critical Technology Elements (CTEs) and verified that CTEs were at least TRL-4
- STP project team updated Technology Maturation Plan ➔focus on integrated testing in a relevant environment with a range of bounding simulants.
- Conducted testing and integrated testing in conjunction with preliminary and final design completion. Integrated Testing of prototypical system with a range of bounding simulants.
- STP conducted a second TRA self-assessment to verify that all technology elements were at least at TRL-6 ➔ built documentation basis for TRL self-determination
- DOE team identified 5 CTE's; and verified that all were at least at TRL-6

U.S. DEPARTMENT OF ENERGY

CH2MHILL Plateau Remediation Company 20

Current Project Status

- **Project has received CD-2/3 approval**
  (approved to procure and complete construction)
- **KW Annex (STSC loading facility) is under construction**
- **Procurement of sludge handling equipment is underway**
- **Sludge Test Facility is being used to further optimize design**
  - Improved control system architecture to simplify installation in the contaminated basin
- **All production sludge handling equipment will be installed in the test facility as part of acceptance and initial cold commissioning**
- **Operators are using test facility to develop procedures and begin training and qualification**

U.S. DEPARTMENT OF ENERGY

CH2MHILL Plateau Remediation Company 21

Bottom Line

- **DOE's TRA process establishes <u>requirements</u> and <u>expectations</u> for technology maturity during the design phase and critical decision approvals**
  - Replaces individual expert judgment with a team based questionnaire methodology with expected maturity demonstrated at each project stage
- **DOE's adoption of the structured TRA evaluation process has driven more extensive pilot and scale testing than the previous norms for DOE projects**
  - Requires time and money early in the design cycle
  - Upfront costs are off set by significant savings in startup and initial production
- **DOE's adoption of the structured TRA process results in a more rigorous approach to component and system testing**
  - Demonstrate to an independent team that adequate testing has been done at each stage of design

U.S. DEPARTMENT OF ENERGY

CH2MHILL Plateau Remediation Company 22